The Day the
WHISTLE BLEW

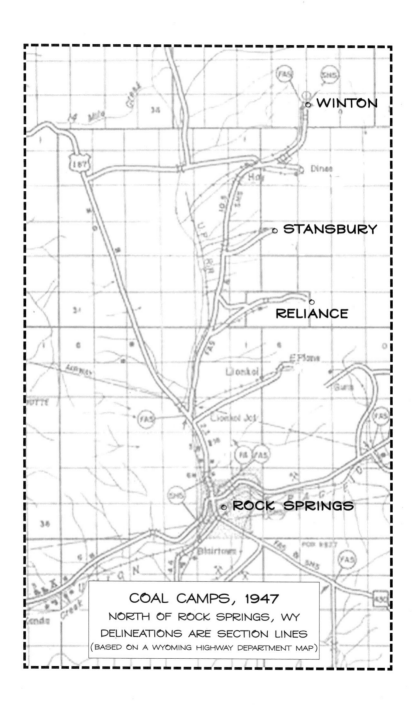

COAL CAMPS, 1947
NORTH OF ROCK SPRINGS, WY
DELINEATIONS ARE SECTION LINES
(BASED ON A WYOMING HIGHWAY DEPARTMENT MAP)

The Day the
WHISTLE BLEW

THE LIFE & DEATH OF
THE STANSBURY COAL CAMP

Marilyn Nesbit Wood

HIGH PLAINS PRESS

LIBRARY OF CONGRESS CATALOGING-IN-PUBLICATION DATA

Wood, Marilyn Nesbit
The day the whistle blew : the life & death of the Stansbury coal camp / Marilyn Nesbit Wood.
    pages cm
 ISBN 978-1-937147-08-2 (pbk. : alk. paper)
 1. Coal miners--Wyoming--History. 2. Coal miners--Wyoming--Social life and customs. 3. Coal mines and mining--Wyoming--History. I. Title.
 HD8039.M62W88 2014
 331.7'6223340978785--dc23
                    2014020266

HIGH PLAINS PRESS
403 CASSA ROAD
GLENDO, WY 82213

WWW.HIGHPLAINSPRESS.COM
ORDERS & CATALOGS: 1-800-552-7819

FIRST PRINTING

10 9 8 7 6 5 4 3 2 1

Manufactured in the United States of America

In memory of those I loved who have passed away:
my mother, father, and brother Jimmy,
Aunt Ann and Uncle Zeke, Uncle George and Aunt Edna Copyak,
and Dr. John P. Muir.

And to my children, brother Johnny, and Aunt Kay Croen
whose love inspires and sustains me always.

And to Margot Joy, my dear friend,
who was the first and the last person to read these pages
in manuscript form and who helped me selflessly.

And special thanks to my publisher and friend, Nancy Curtis,
who lent her careful eye and sharp editing skills to my book
and helped me immeasurably.

All the incidents in this book actually happened. Any errors of fact or memory are mine. I have recounted the events as accurately as I could, relying upon personal memory, the memories of family and friends, and published documents and accounts. I also referred to my own diaries and letters of the era. One name has been changed in consideration of the privacy of the individual.

•• MARILYN NESBIT WOOD

CONTENTS

It's hard to forget a place
that gave you so much to remember.

-- UNKNOWN

CHAPTER ONE

REUNION

THE COAL CAMP reunion was over, but it left me with more longing than if I'd stayed home. I'd been warmly greeted at the reunion headquarters hotel in Rock Springs, Wyoming, by other "coal camp kids," now gray-haired and bifocaled "seniors" in their sixties and seventies. We'd shared photos and visited into the night; then the next morning we all climbed onto the buses to tour the old mining camps, company towns established by the Union Pacific Coal Company. At Winton and Reliance, we walked around looking for familiar sites, recalling things that had happened there. But Stansbury, the abandoned town where my memories lived, was blocked off by NO TRESPASSING signs. We were told the new owners of Stansbury would not give permission for us to cross the barriers.

Back at the hotel, I said goodbye to those people I love so much and got into my car. I drove down Dewar Drive, turned off at the Elk Street exit, and headed north. I had but one thing in mind and that was to go out to Stansbury and somehow, someway, cross that barrier to revisit where I had come from.

As my car sailed farther out of town, I saw a scraggly bush growing up here and there but, for the most part, the land was the kind you have to learn to love. The butte to the north wasn't important enough to have a name, but I recognized its low, humped-back shape, rising from the dirt like a beached whale. How many times had I passed this landmark as a girl? And each time my family had been waiting for me at Stansbury, at home.

o x o x o

Five miles north of Rock Springs, I turned off the main highway onto the road leading toward Reliance, the coal camp where kids from Winton and Stansbury were bused to attend junior and senior high. Just before the Reliance exit, I took the fork leading to Stansbury, another three miles farther down the road. I slowed on the dilapidated highway, which had once been so heavily traveled. The potholes made it difficult to drive more than twenty miles an hour. No cars came down the road toward me, nor did I see any in my rearview mirror. The strong desert winds whistled all around, stirring up sand and dirt. At times a gust made it hard to see the road ahead of me. Huge tumbleweeds rolled across the plains onto the highway. Nothing was going to stop them for miles. I felt as if I were traveling into some off-limits nuclear testing site. I looked for a place to turn around to go back to Rock Springs. Whatever made me think this was a good idea?

I knew once there had been a sign marking the route to Stansbury. Years ago, the local 4-H Club had made a big wooden sign which they painted white and then painstakingly printed in big, bold, green letters: STANSBURY. The jagged post that once held it was still there, marking the site, but the sign was gone. Slowly I drove down what was left of the road leading to Stansbury.

○ × ○ × ○

In the distance I saw a barricade across the road with a heavy chain, supported by two rusted steel poles holding a threatening NO TRESPASS-ING sign swinging in the wind. Now the eerie feeling I'd had earlier only heightened. Tall sagebrush encased both sides of the highway, making it impossible to even see the drop-off where a creek had once run or to see a safe place to turn around. When I reached the barricade, I left the car running while I got out to see if I could unhook the chain and get through. I watched each step I took, fearful of rattlesnakes. Doubt and helplessness made my heart rise in my throat. What would I do if I needed help to get out of the situation I had gotten myself into?

A rumble caused me to look up, and I saw a white pickup truck traveling rapidly straight toward me. It came to a halt directly in front of me, on the other side of the no-trespassing sign. The driver jumped out of his truck, slammed the door, and angrily shouted as he stomped toward me. "What the hell are you doing out here, you crazy son of a bitch? Can't you read? Get your ass back in your damn car and get the hell out of here." I froze.

He was a husky, elderly fellow wearing bib overalls, a red plaid shirt, and a hardhat. I strained to see his face under the hat. It looked somewhat familiar. I yelled back over the wind, "I used to live in that mining town over there! I haven't come out here for a long time!" Determined not to be intimidated by his confrontational manner, I persisted as I held onto the heavy chain in front of me and yelled, "Can you unlock this chain so I can drive up this road a little farther and just look around the camp?"

Now he was standing right in front of me, eyes blazing. "Who the hell are you anyway?" he asked.

"Marilyn, Marilyn Wood . . . used to be Marilyn Nesbit. I grew up here. I just want to see what Stansbury looks like now!" My eyes teared up, and I hoped he would think it was the dust and wind.

He stared at me for a few seconds and then asked in a softer tone, "You're Johnny's daughter, aren't you?"

It had been such a long time since I had been referred to as Johnny's daughter that I stood speechless. Then he continued. "I remember you when you were just a little girl. I worked for your dad. We lived right across the street from your family."

"Who are you?" I asked still unable to put a name to his face.

"Joe Bozner. Don't you remember me?" he queried.

The tears escaped my eyes and ran down my cheeks. "Of course I remember, but I never dreamed I'd see anybody I knew out here today, much less you."

"Come on," he said as he unlocked the chain. "Pull your car over and turn it off. Better lock it. We'll go up to Stansbury in my truck."

When I was in his truck, he said, "You won't even recognize the place anymore. Bet you won't even know where your house used to be. I'm telling you, all that's left is the mine portal and a few old sheds, and then nothing for as far as the eye can see." He paused for a moment, "I'm retired now. Just took this watchman job to fill my time and make a little extra money. You know, something to keep me busy. I'm not one of those people who retires and just sits around the house day in and day out. Zocka, my wife—remember her?—and I live over in the little house by the railroad tracks just before you get to Reliance. You'll have to stop over to see her on your way back. She'd sure like to see you again." As we drove toward the abandoned mining town I'd once called home, I peppered him with questions.

Finally, I took a deep breath, and he took advantage of the pause. "Always wondered what happened to your family. Tell me about your mother and brothers. Is your mother still. . . ."

I knew he was afraid Momma had died. "Oh, they're all fine. Momma lives in Phoenix now. My brother Johnny lives in Phoenix, too, and my brother Jimmy is in San Francisco. I'm in Laramie."

"Do they ever come back to Rock Springs?" he asked.

"Oh, from time to time," I replied. "But for some reason they never have wanted to see Stansbury again." There was a silence as we both considered what draws some people toward the past and holds others away.

"I came for the coal camp reunion, Joe. But I was told that no one was allowed out here. Why's that? What harm could be done by people just looking around?" I asked.

"Hell, it's too dangerous! Liability," he exclaimed. "Rattlesnakes everywhere. After the company sold all the houses, the buyers had to move them out. Nothing but a bunch of empty foundations were left standing. The next thing we knew, a bunch of runaway kids—hippies or people down on their luck, I guess you'd say—were actually living in the foundations. Some were even living inside the entrance of the mine itself. People thought strange things were happening out here, illegal things, and maybe some of that was true and some not. Cops were always being called to check on what was going on. You know, those damn kids might of gotten killed if they went too deep in the mine or got into one of those air shafts scattered all over these hills and ran into black damp. So, UP bulldozed the whole damned place, chained off the camp, and hired watchmen, retired guys like me, to keep an eye on things out here as long as they owned it. That way no one would get hurt, and UP wouldn't get sued."

As we topped the hill, I could feel the blood drain almost down to my toes. There it was—Stansbury! All I could see was a barren piece of land with a few trees standing here and there. All the houses and buildings were gone. Joe brought the truck to a stop, and we got out.

"God, Joe, I can't believe what I am seeing." Everyone dreams of returning to where they started, to their hometown, completing the circle back to where they began. But there was nothing here. My hometown had been wiped off the face of the earth. "Oh, Joe!" I couldn't hide the tears from Joe anymore and reached into my purse for a tissue.

"Ah, come on now, don't cry," he said, as he reached over and put his hand on my shoulder. "We both have a lot of good memories living here. Things can't stay the same. Everything changes. Just think about how it used to be."

He quickly turned away and looked out over the space. "Bet you can't even tell where the general store was, or the post office, let alone your house. But I can show you where your house was, since I lived right across the street from you folks. I've been around here long enough to remember just where things used to be."

In the distance I saw the entrance to the mine and two newer modular warehouses. A rusted old Joy Loader partially buried in the sand to the side of the portal guarded the entrance of the mine. It could have been the very piece of machinery Daddy tried to dive under the day of the cave-in. The entrance of the mine looked so much smaller now than I remembered. I found myself thinking about the time when my friends and I hauled orange crates to form a stand right outside that entrance. We made a can full of pennies selling Kool-Aid to the miners as they came out of the mine at the end of day shift.

I cleared the lump in my throat and asked, "Joe, could we walk down into the mine a little way? I want to see how it is."

"Yeah, but just a little ways," he answered hesitantly. "Gotta be real careful, you know."

As we entered the mine, a cold, damp breeze came up from way down deep, and a familiar smell permeated the air around us: mine water. The mine was dark, much darker and much colder than I ever remembered. Joe flipped on the lamp on his hardhat as we continued walking farther and farther down into the mine. The light helped guide our way straight ahead, but it was still jet black around us.

"How could Daddy have worked in a place like this?" I muttered.

Joe was quick to defend, "You know, strange as it seems, a miner gets used to working underground. I loved it, and I bet your dad did too. Farther down it's like a city all its own where the temperature never changes. I guess it's all what you are used to. Sure, mining's dangerous. Every damn move you make is dangerous. But you just go to work each day hoping to hell today isn't the day the good Lord means for you to get yours."

"When Daddy left the house to go to work each day I didn't have any idea how dangerous the mine was. I guess I felt he was so big and strong that nothing could ever happen to him."

As we walked along, I could think of nothing but how my father had worked this very mine for nearly twenty years, many times more than his eight-hour shift. When he came out of the mine each day, his face and overalls were covered with black coal dust, and a ridge in his forehead revealed where his hardhat, heavy with the battery-operated lamp, had rested all day. He was unrecognizable until he came out of the bathhouse. People said he was as handsome as a movie star, and he always had a smile on his face no matter what had transpired on the job or what dangers he had endured underground.

"See that area over there?" Joe interjected. "That is where I found your dad's old mine desk. Ya see, when UP closed the mine for good, they sealed off the entrance with cement. The mine soon filled up with water. Years later another company bought the mine from the UP. They opened up the entrance again and drained out all the mine water. I loved working here, so I hired on again when the new guys took over. I was working the day they drained it, and I found your dad's desk come floating up." Joe's gaze rested for a few long seconds on the spot where he'd last seen the desk. I knew he was remembering more. "Well, come on now, let's get out of here. It's getting too cold for me. Besides, we've gone as far as we can go. I'll drive you around the camp." Together we walked out of the mine, each using one hand to help the other along on the uneven footing.

"Joe, you say you liked working for UP?" I asked when we reached the daylight.

"Oh, I did and I didn't, if you know what I mean. It was the hand that fed us. UP's main concern was production. Overall they were a pretty good outfit. I got a good retirement out of them. The black lung settlement was a godsend."

He paused, concentrating on stepping around some rubble, and then continued, "What the hell was the name of that doctor who went into the mine that night? I'll never forget that man. That took guts for him to go down in the mine under those conditions. I heard that was the first time he had ever been underground. Wonder where he is now?"

"It was Dr. John Muir, Joe," I replied. "He died when he was only forty-two years old."

"Ah, I hate to hear that," he replied. "Sure was a wonderful man and a damn good doctor! Like I said, it took a lot of guts for him to go down in that mine when he'd never been underground before. Risked his own life, ya know, crawling through that narrow tunnel the guys had dug with their bare hands, just to see if anyone made it through those two big cave-ins. Even the miners were afraid! They were down there, but they were afraid. Rock kept coming down all over the place during that rescue."

We climbed back into his old pickup and drove over the rocks and weeds until we came to a place where one lone tree stood. "Recognize that tree?" he asked. "That's the willow your dad planted after he put up that big white picket fence . . . remember? Don't know how it managed to survive all the bulldozers and people coming out here to take whatever they could, even the trees, after they moved all the houses out. Some people called them scavengers, but that was too nice a word for them."

"Wait," I exclaimed. "Let me get out for a minute, Joe. See that pile of cement over there? Let me see if I can find where we kids put our palm prints in the fresh cement Daddy poured to make our sidewalk."

"Oh, I don't think you'll ever find anything like that in all that rubble," he chuckled.

"Just give me a minute," I begged. With Joe watching over me, we got out of the truck and walked to where our house had stood and slowly I picked through the broken pieces of concrete, looking for remnants of that day so long ago.

"Here, let me see if I can help you," he grunted. I could see he was patronizing me. Together we pushed and pulled until I found one piece of cement that had "Jimmy" written in cursive. My brother Jimmy.

"See, Joe! The minute I saw those pieces of concrete, I had a feeling I would find some trace of that sidewalk. Boy, did we get in trouble that day for scribbling in the wet concrete!"

I suppose Joe knew what was going through my mind. Sitting among the rubble, we both leaned toward each other and wrapped our arms around one another without saying a word, sitting silently with only the sound of the wind whistling around us. This time not a tear came to my eye. I was so emotionally moved that I couldn't even cry though I could feel my heart breaking.

Suddenly Joe jerked away and looked toward the sky as he said, "Come on now, kid, I've got more places to show you before my shift is over."

"Wait, Joe! Would it be okay if I just sit here for a while by myself?" I asked. He paused to consider my request. "Sure. The ground is flat enough around here, and I don't see any snakes. You should be okay if you stay right here. I'll come back after I make sure all the gates are locked and finish up a little paperwork. Probably take a half hour. But I know where you are so I'll keep an eye on you. Just be careful! Wave if you need me."

"Thanks Joe," I replied. He slowly climbed back into his old truck and drove away. I soon found myself slipping into another place in time and remembering things as if it were yesterday. It was then that I decided the minute I got home, I had to write the story of my family's life in Stansbury. Too much had happened here to let it be forgotten.

After a while, Joe returned and drove me back to where my car was parked. He was even nice enough to turn my car around for me so I could head straight out on the highway without getting stuck. I couldn't thank him enough as we said our goodbyes.

All the while I was growing up, I'd thought there would always be a Stansbury, just like there would always be a Rock Springs. Never in my wildest dreams did I think the day would come when Stansbury would be completely obliterated.

CHAPTER TWO
FRONTIER OF HOPE

BOOKS AND MUSEUMS are full of stories of the settlement of the West, often seen through a soft focus lens. Mountain men trapped, hunted, and panned for gold in pristine mountains and streams, then joined together for raucous rendezvous with welcoming Native Americans. Emigrants moved families westward via wagons and handcarts looking for a fresh start on virgin soil. Ranchers trailed cattle to fertile grasslands facing down the whims of weather and rustlers. But the stories of the men who mined underground coal are harder to cast under a spell.

Underground coal miners and their families traveled to their destinations by ship or train and set up homes in areas often just as unsettled as those encountered by other pioneers. Hundreds came from all over the world to the desolate landscapes and carved towns and camps along the Union Pacific railroad.

Towns like Rock Springs, Superior, McGath/Winton, Dines, Eplane, Quealy, Reliance, and Stansbury were born when miners brought coal—and hope—to the surface. Each coal town was a unique village. While having some similarities, they each reflected the diverse ethnic groups, cultures, interests, talents, and skills of the particular people living there.

The coal towns prided themselves on being melting pots for immigrants of all nations. However, in September 1885, a Rock Springs coal settlement of Chinese miners was burned out in what started as a labor dispute between white and Chinese miners. Over seventy-five Chinese homes were burned and at least twenty-eight Chinese were killed, perhaps as many as forty. The number is not known because families ran from their burning homes into the night, and some people were never seen again. Federal troops arrived to establish the peace. So some ethnic differences were not always accepted readily.

A few of the towns thrived and continue to exist. However many townsites, including my own hometown of Stansbury, returned to sage-brush and wild flowers which now cover the gigantic coal seams hunkering silently underground.

○ ✕ ○ ✕ ○

I am a coal miner's daughter. I heard the story of my father's coming to the Wyoming coal fields each holiday as we ate turkey, creamy mashed potatoes, stuffing, and pie. His story became nearly as much a part of me as the place itself.

My father, Johnny Nesbit, a young man desperate to find work in 1934, never wanted to come to the coal fields of Wyoming. He had grown up seeing his father endure hardships as a coal miner working in the Peabody Mine in Clinton, Indiana, and he'd vowed never to be a miner. He found it painful to even think of working eight hours a day in a cold, dark, damp, and often dangerous, mine. He had absolutely no desire to work day in and day out, breathing in the coal dust created by the underground blasting, digging, and machinery, hoping that the carbide lamp on his hardhat kept burning, eating lunch out of a tin pail and knowing that every bite would smell and taste like coal, and ending each shift covered in so much grime and black coal dust that the only way to recognize a man was by his eyes. But in the 1930s, our country was in a depression, jobs were almost non-existent, and Johnny was one of nine children, six girls and three boys.

Even though his father had steady work at the Indiana mine, he could only afford a one-bedroom house for his family, who slept on the living room floor in bedrolls. It took more than he alone could earn to provide even the bare necessities for the family.

When the worked-out Peabody Mine closed its doors for good, Johnny, who had just completed the tenth grade, dropped out of school and, with his older brothers, looked for work to help support the family.

However, the entire community was depressed. Families had no money to relocate and no place to go. So when the owner of the local fruit market placed a "part-time help wanted" sign in his store window,

Johnny quickly applied for the position and was happy to be hired. At the end of each week, he picked up his check, cashed it at the bank, and took the money to his mother, who reluctantly took the cash. Johnny often recalled how his mother, with tears in her eye, thanked him in her Scottish burr as she slowly counted out the money she needed and handed him back a few dollars for his spending money. Johnny continued to help support his parents, from each paycheck, all their lives.

Johnny loved working in that fruit market and looked forward to going to work each day, stocking fresh fruits and vegetables and waiting on customers. He dreamed of having a market of his own like that.

He saved the money his mother turned back to him for a day when he had enough to leave Clinton to look for a better job. The family scrimped and all pitched in to get by. Johnny, with his tall, athletic build, and dark wavy hair, drew the attention of neighborhood girls and became known for his Saturday night dancing.

○ X ○ X ○

Johnny's older sister, Jenny, was the first to leave home. She met and married Bob Wilson, who was a supervisor at Peabody Coal. He worked for Peabody until the mine closed. He had excellent mining skills and so was contacted by other mining companies with job offers.

My Aunt Jenny often told me about the evening a stranger approached their front door. Bob explained that the man was a Union Pacific official he'd been talking with earlier that day about possible work at a coal mine in Wyoming. Bob knew he had to find a job somewhere and was eager to hear what the official had to say. The meeting that night changed the course of events for the family.

Bob decided to take UP's offer to travel to Wyoming, all expenses paid, to see the mine before making a final decision.

A week later Bob was met at the Rock Springs, Wyoming, train depot by UP officials who drove him out to a mine at Winton. He was immediately struck by the sharp contrast between the high desert of southwestern Wyoming, lacking both trees and water, and his home in Indiana. As the men traveled ten miles north of Rock Springs along the winding highway to Winton, all he saw out the car windows was wide open sagebrush flats.

As they drove into town, where the Winton mine had begun operation in 1921, he saw a hodge-podge of buildings around the mine office. Scattered around the surrounding hills and along the main street, were older, well-maintained company houses with manicured yards. The main street was the only paved street in the entire town.

The officials took him directly to the mine office where he met the superintendent and began a tour of the surface and underground operations. The company took pride in its safety record, offered excellent pay, and provided adequate company housing. He liked the attitude of the miners he met along the tour. He was amazed to see the steady stream of coal locomotives pulling dozens of cars down the winding railroad tracks to and from Winton, hauling huge amounts of coal.

At the conclusion of the tour, the superintendent made Bob a very tempting job offer. However, before he signed his name on the dotted line, he knew he must convince Jenny to become part of this booming new frontier.

When he got back to Indiana, he couldn't stop talking about all he had seen on his trip to Wyoming. Aunt Jenny could not disappoint him and reluctantly agreed to leave her family and the beautiful Indiana town she loved so much to follow her husband.

Johnny had tears in his eyes the day he watched the heavily loaded truck carrying Bob and Jenny's possessions drive away from the Indiana house. He wondered if or when he would ever see his older sister again. She promised to write every day as their vehicle pulled away, following the Wabash until it faded out of sight.

o x o x o

And write she did, every day, just as she'd promised. Each time a letter arrived, Johnny and his family gathered in the living room after supper, listening closely as their mother read Jenny's letter. They always started out by telling how she missed everyone and what a stark difference the barren prairie of Wyoming was compared to the green, rolling hills of Indiana. She recounted how difficult she found adjusting to the rattlesnakes outside the house and scorpions everywhere, in and out of the house. But, with her Scottish determination, she set about making Winton her home.

In letters that still survive she tells how isolated she felt. Her husband, who held the position of mine boss, or "company man," was on call at the mine, often being called back to the mine after working his eight-hour daily shift. He made $350 a month, regardless of how many hours he spent at the mine. At that time, $20 bought a full box of groceries and $1.25 bought a rump roast big enough to feed six people.

Bob had many men working under him—foremen and shift workers. Unit foremen were also considered company men. They too were paid a flat monthly salary with no over-time pay and were on call-out twenty-four hours a day. The shift workers were paid an hourly wage much lower than that of their supervisors but were eligible for time-and-a-half for working more than forty hours a week and double-time on holidays. With the overtime pay, a shift worker's salary could be considerably higher than that of the company man. However, there was one distinct difference: a company man's job was full time; whereas the shift workers were considered part-time employees, subject to layoffs and strikes.

As time went on, Jenny's letters took on a different tone as she settled into the community and met other wives. The miners' wives had the responsibility of running their households: tending children, cooking, baking their own bread and pastries, and cleaning. In addition, they tackled the family laundry using primitive conventional washing machines and water heated on the kitchen coal stove. The wives were plumbers, carpenters, peacemakers, teachers, beauticians, and sometimes even dentists and physicians.

The women also made their own entertainment. They formed card clubs which met in homes and required their Sunday clothes and best manners. Each hostess prided herself on serving delicious refreshments. Alcoholic beverages were never served because it was considered inappropriate for a woman to drink outside the company of her husband.

When they had tallied their scores, the hostess handed out homemade prizes. They shared stories about their families, talked about the latest movies or radio programs, and engaged in camp gossip, especially suspected clandestine love affairs. However, they never mentioned their husbands' jobs or accidents or their panic at the unscheduled sound of the mine whistle. Each wife feared for her husband when he was in the mine, but to voice those fears might cause them to become realities. Superstitious? Yes, very much so.

The wives also joined the local Extension Homemakers' organizations which held monthly meetings in the community hall. The organization sponsored fund-raising potluck dinners and cakewalks for many worthwhile community activities and taught homemaking skills.

The women came from varied ethnic backgrounds, thus sometimes creating a language barrier. But the "sameness" of their lives, the sharing of isolation, and the underlying danger of a miner's work unified them. A familiar phrase was uttered by many miners' wives as the husband left the house to go to work: "Now you come home to me."

○ × ○ × ○

But Jenny still longed for her family. Finally, Bob suggested that she write and ask her brothers if any of them wanted to apply for work in the coal camps. Before long they sent Johnny a train ticket to come to Wyoming. Jenny was thrilled.

CHAPTER THREE
WYOMING BOUND

F OR YEARS DADDY related the story of the hot summer after-noon in July 1934 when he arrived at the Rock Springs train depot. He was twenty years old, feeling lost and bewildered, never having been out of Indiana before, when he stepped off the train onto the depot platform. His first impulse was to hitch a ride on the next train leaving Rock Springs heading east and go home. But he knew that wasn't an option. Having given most of his last paycheck from his old job at the market to his mother, he didn't have much money in his pocket. Bob and his sister Jenny were there to meet him. Jenny spotted him right away in the crowd that day with his tan trousers, white shirt, dark jacket with matching vest, and polished dress shoes. She ran up to him, throwing her arms around his neck and clinging to him while tears streamed down her cheeks. Softly she whispered in his ear in her slight Scottish burr how glad she was to see him.

The ten-mile drive from Rock Springs out to the mine camp of Winton proved to be even worse than he'd imagined. For a few moments he sat quietly with his thoughts while Jenny jabbered incessantly. He hated the bleakness and ugliness of this barren countryside and wondered what he had gotten himself into. After seeing Rock Springs, he couldn't begin to even imagine how bad Winton must be. With each passing mile, he became more convinced that he had made one big mistake by coming to Wyoming to work in—of all places—a coal mine. As he watched the passing trains struggling along the tracks carrying heaping cars of coal from the three mining camps north of Rock Springs—Reliance, Winton, and Dines—his hand automatically felt for the few

My father, Johnny Nesbit, was a dapper, athletic young man when he arrived in Winton to begin his first job in a coal mine.

bills in his pants pocket. He didn't have the money to escape this god-forsaken, desolate country. The only way out might be to hop a coal train leaving the coal camp for Rock Springs, and then catch a freight train headed back east.

The wind was blowing hard, kicking up dirt and sand until he could feel sand coating his moist lips as he looked out the open window of the car. A sea of rolling tumbleweeds blew across the highway and bounced across the plains on both sides of the highway. His heart plummeted when he finally caught sight of Winton off in the distance. He thought it looked like the end of the world. However, as they neared Winton, he saw trees, something he hadn't seen anywhere on the landscape since they had left Rock Springs. Jenny, seeing the look on his face, reached over and put her arm around his shoulder in a gesture that was both pos-sessive and protective, reassuring him that it wasn't as bad as it looked.

As they drove into Winton itself, Johnny was shocked. The old mining town looked like an oasis in the desert with well-maintained older houses and manicured yards filled with flowers and trees.

Jenny insisted upon taking her brother's photograph in the driveway. I've studied every detail of that photo so many times I've almost worn the sepia color right off the paper. The man who was to become my daddy looks so young and innocent, so debonair wearing his sporty dress clothes and his sweet expression.

The smell of beef pot roast baking in the coal stove oven filled the air as they opened the front door of the house. The house was spotless inside and looked comfortable with new overstuffed furniture, drapes at the windows, and glistening hardwood floors. The dining room table was set with the china and silverware that Johnny learned were wedding gifts. On the table sat a large glass vase with fresh flowers from the yard. Jenny walked him back to the bedroom she had made up especially for him where she'd spread a new set of work clothes and socks across the bed alongside a box containing a new pair of hard-toed work boots.

For the first time in his life Johnny had his very own bedroom, one of three in Jenny's house. The room had a closet, a dresser, and a framed picture of his parents beneath the lamp on the nightstand. Johnny draped his arm around his sister and gave her a squeeze.

During dinner all Bob could talk about was the mine and the job openings that were available. Jenny interrupted with comments and questions about the family back home whenever she got the chance.

After they finished dinner and straightened the kitchen, they stepped onto the front porch and Bob began to set hoses to water the yard. Jenny sat in one of the wooden rockers on the porch as Johnny slipped off his socks and shoes and walked across the damp grass admiring the yard.

Later when the three of them walked around town, Johnny felt as if he'd stepped back in time. Old, wooden-framed company houses scattered up the sprawling hills from the single paved street leading through camp. Black smoke from cookstoves slowly rose from the chimneys. Outhouses, coal heaps, and clotheslines filled backyards. The smell of sulfur water from the mine permeated the air.

As they approached the far end of the street, Jenny pointed out the UP community hall, bar, pool hall, clinic, the UP general store, and the post office. Across the street was the elementary schoolhouse, located on top of a hill overlooking the town. Jenny explained that the school kids crossed an old wooden bridge built over a creek filled with runoff water flowing from the mine. During heavy rain, this creek sometimes flooded over the bridge making it impossible to cross, requiring a school holiday. Johnny was amazed at the obvious pride the residents took in their community. Bob explained that UP, in an effort to encourage the camps to become more home-like, ran a contest each year offering a $10 prize for the best yards and gardens. The cash award compensated for some of the extra time and work, but the prestige was even more valued.

Jenny pointed out the home of Frank Franch and his family which consistently won prizes with its extraordinary flower garden, flourishing in defiance of the elements. Johnny admired the beautiful hedges of petunias, bachelor buttons, and well-trained sweet peas. In the center of the lawn, the ground sloped downward to a lily pond with goldfish and water plants. Tulips, asters, hardy zinnias, and dahlias flanked the pond.

Down the street, they stopped by the house of one of Winton's favorite doctors. Jenny explained how he had built a small cement pool in his yard for his children. The pool, about three feet deep and ten feet wide, graced a yard filled with flowers and trees.

When they got home, Jenny assured Johnny again that once he got to know the people and got involved in the activities, he'd find Winton wasn't such a bad place after all.

Years later when Daddy told us the story his eyes softened as he recalled the families spending time outdoors, sitting on their front porches visiting with neighbors or tending their yards, calling hello to passersby. On this, his first evening in Winton, he was surprised at how curious the residents were about the young stranger at Bob and Jenny's home.

Later that evening as they got ready for bed, Bob told Johnny they would be getting up early the next morning in order to report to the mine at seven.

Daddy often told me how hard it was to fall asleep that night. All kinds of thoughts ran through his mind as he lay looking up at the stars through the window next to his bed. There was no getting around it: he missed his family and home and felt trapped knowing there was no turning back. He desperately needed a job, but the thought of working underground made it hard for him to breathe.

○ ✕ ○ ✕ ○

A little before sunrise the next morning, Winton came alive. The headlights of vehicles carrying miners from Rock Springs lit up the street. Dim light flowed out kitchen windows in practically every house in camp. Steam engines, pulling heavily loaded coal cars, thundered down the railroad tracks leading out of Winton. In the dawn of early morning, a bevy of miners kissed their wives goodbye before leaving their homes to work the day shift. With lunch buckets swinging in their hands, they walked in silence down the faintly lit dirt streets toward the bathhouse.

Jenny scurried around putting the finishing touches on the lunches and cleaning up after the hot breakfast of bacon and eggs. The round metal lunch buckets had two compartments: the top held the lunch, and the bottom was filled with drinking water. A couple of sandwiches, chips, pie, and a piece of fruit usually sat in the top compartment. Each miner also packed a Thermos of coffee. Bob kissed Jenny goodbye before he and Johnny grabbed their things and walked out the back door into the early morning air headed to work.

The first person the miners saw when they entered the bathhouse was one of the fire bosses, seated at an old wooden desk, writing his morning inspection report. Usually three fire bosses went down into the mine before the morning shift to inspect each of the different seams of the mine. They did a safety check for gases, roof conditions, coal movement, or other suspicious changes, then turned over their daily report to the mine foreman to review before he allowed any work crews to enter the mine.

○ ✕ ○ ✕ ○

Promptly at seven o'clock, the mine whistle blew loudly signaling the beginning of the morning shift. Johnny reported to the mine office and signed up to become an employee of Union Pacific Coal. He filled out the personnel paperwork and learned that UP paid every two weeks. An office worker shoved a handbook toward him and told him it was mandatory to read the handbook in its entirety and sign a paper attesting to this fact. The title on the handbook read: "Rules and Regulations for the Government of All Employees of UP Coal in Accordance with the Provisions of the Law of the State of Wyoming Regulating Coal Mines."

Johnny looked at the list of all the current position openings posted on the office wall—electrician paid the most and didn't involve swinging a shovel all day. Although he had never worked with electricity, he recalled Bob's earlier advice and signed up for the electrician apprentice program. Next he went to the bathhouse where he changed into work clothes and safety boots.

The bathhouse change room was one large room divided into two sections. The large "gang" shower area ran the length of the building with no dividing partitions. This area had tiled walls, a multitude of showerheads running across the ceiling, and a cement floor with drains. Directly across from the shower area, a series of wooden benches sat in rows. A network of chained trolley baskets, set approximately two feet apart, swung from the ceiling above the benches. Each basket could be lowered or hoisted to the ceiling via a heavy chain. The miner padlocked it into place high in the air for security purposes.

After changing into work clothes, each miner stored his clothes and personal items in the basket while he was down in the mine. At the end of a shift, he picked up a clean bath towel the minute he entered the bathhouse. He lowered his basket down from the ceiling, took off his work clothes, showered, and changed into his street clothes. He hung his work clothes on a hanger on the outside of the basket, hoisted the basket back up to the ceiling, and locked the combination padlock before signing out for the day. Miners took their work clothes home to be laundered at the end of each week; they worked in the same dirty clothes all week long.

The mine women made no secret that washing the filthy mine clothes was an onerous task. Since coal dust permeated the miners' clothes, the wives soaked the clothes in salt water in a tub in small laundry rooms or kitchens and scrubbed them on scrub boards. Then they put them in the hot soapy water of a conventional washing machine. Some of the miners' wives even ironed the work clothes, so they would feel softer. When they did this, the ironing board cover was black when they finished ironing, and it had to be laundered before being used again.

○ X ○ X ○

Ready for work, Johnny went to the front office of the bathhouse where he was issued a hardhat and lamp, and taught how to hang a methane gas tester from his belt. Finally, he was issued a brass identification badge the size of a quarter with his specific employee number engraved on it. This identification number would stay with him as long as he worked for Union Pacific. Johnny's number was 701. Each miner kept this badge in his left shirt pocket as he walked out of the bathhouse to begin the day's work. At the end of the shift he hung the badge on the "mine badge board," whether he'd worked underground or on the surface. Before leaving the bathhouse for the day, the unit foreman checked the badge board to be sure that every miner in his unit was accounted for.

After Johnny picked up his badge, the bathhouse boss took him to meet the electrical crew. The foreman handed him a stack of manuals to take home to study for his journeyman electrician license. The foreman then held a tool pouch out to him and said, "guard it with your life, Johnny, because this is the only one Uncle Pete gives you. If something happens to it, you have to pay to replace it yourself." Johnny didn't question who this Uncle Pete was, but before the day was out he realized that Uncle Pete was in charge of most everything in the mine as Uncle Pete was the miners' nickname for the all-encompassing Union Pacific.

Johnny and the foreman hurried to catch the "mantrip" that was about to go down into the mine. "We'll take this one down and catch the next one coming back up," the foreman said as they ran and jumped into one of the cars.

"Ever been underground before, Johnny?" the foreman yelled as they ran toward the mantrip.

"No, sir." Johnny replied apprehensively.

About twenty men squeezed onto the yellow benches in each of the ten cars of the mantrip, which was pulled along on a trolley-like system. The miners sat in a leaning back position to make the vehicle as low profile as possible. The lead car was connected by a pole to electrical lines running along the ceiling of the mine. The switchman flipped the brake, and the cars started rolling down the one set of tracks leading into the mine, the same track used by the coal cars. Johnny could feel

Johnny Nesbit, my father, (back row center) was well liked and had many friends. Here he poses with friends in Clinton, Indiana.

his heart race as they entered the mine portal and headed into the darkness below.

He had never envisioned himself doing this. The dark shaft, which seemed no wider than the beam from his headlamp, took on a musky smell. For many days this first trip underground haunted him: the smell of dampness and sulfur, the bleakness, the sinister cracking sounds of coal shifting, the water dripping from the roof in places, and rats scurrying about. The sound of heavy equipment and blasting could be heard coming from deep down in the mine. He could see why many men turned in their equipment and quit the mine the minute their first shift ended. But he didn't have a choice. He had to stick this out as his family back home was counting on him for financial help.

On that first day, his supervisor introduced him to miners working in each seam. When they reached the underground foreman's office area, he took out a set of blueprints denoting the Winton Mine's wiring system. Step by step he explained how things worked, what had to be maintained, and what Johnny would be expected to do as an apprentice. For safety reasons, electricians always worked in pairs. Johnny was introduced to his partner, who had a look of authority that Johnny came to respect. He was glad to see the light of day when the shift was over.

Routine quickly fell into place on the new job. Johnny made new friends and found being an apprentice electrician was a challenging profession and not just a job. He learned that an electrician played an important role both underground and on the surface of the mine.

o x o x o

During the week, he spent his evenings helping out around the house and studying the journeyman's manuals late into the night. The day he picked up his first check, he went to the Winton UP Store, proudly bought a radio in a shiny cabinet and had it delivered to Jenny and Bob's house as he knew his sister longed for one. Then he put cash in an envelope and mailed it back home to his parents in Indiana.

On Saturday night, he dressed up in the only casual suit he owned and hitched a ride on a coal train leaving Winton to go to a dance at the Playmore Dance Hall in Rock Springs. This became his weekend routine. After midnight, if he wasn't lucky enough to catch a train headed for Winton, he walked the ten miles home in the dark following the railroad tracks. Before many weekends went by, he'd made friends at the dancehall with other young men from Winton who offered him rides to and from the dance in their cars.

It was at the Playmore he first met the young woman who would become his wife. From the moment he saw her, he couldn't take his eyes off her as she danced around the ballroom floor.

How often Daddy told me that she was the most beautiful woman he'd ever seen and how much she resembled Ava Gardner, a famous film star of the times. She had fine features, velvety brown eyes, fantastic legs, and dark hair that had the sheen of a pricey mink coat. She carried herself with elegance and stood out in the crowd dressed in a Chinese-red dress and matching heels. Though intimidated by her beauty, Johnny knew he had to meet her before the band played the next song.

He forced himself to get up the courage to walk across the room to introduce himself: "Hi, I'm Johnny Nesbit. Do you foxtrot?"

"Doesn't everybody?" she responded, laughing as she threw back her head and took his hand to follow him onto the dance floor.

"I haven't seen you here before," he said. "Are you new in town?"

"Oh no. I was born and raised in Rock Springs. But this is the first time I've been to the Playmore."

He liked the way she felt in his arms as they danced, the perfume she wore, her bubbly personality, and the way she followed his dance steps. When the music stopped and they walked off the dance floor, he asked her name.

"My name is Margaret Copyak. Some friends call me Margie, but Momma likes people to call me Margaret."

After that, they danced practically every dance together. Johnny told her about himself. She said this was the first time her father had allowed her and her younger sister Katie to come to the Playmore and that they both worked as secretaries in the office at the UP Store.

"I've got to ask where you got your beautiful dress?" Johnny was so smitten that he was trying hard to make a good impression.

"Oh, my older sister, Ann, is a buyer for a department store in Denver. She sends the latest clothing to me and Katie."

"Well, you look beautiful."

She thanked him for the compliment, and he asked her to dance the next dance. And the next.

At midnight, he blurted out that he'd love to take her and Katie home, but he didn't have a car. Then he asked if she would let him walk them home.

After some negotiating, the three walked along the dimly lit streets leading to the girls' home on Ninth Street which was quite a distance. Before Johnny said goodbye, Margaret agreed to meet him at the Playmore the next Saturday night. With that, he cut between the houses and headed over the hill to the railroad tracks to begin his long walk home. The sound of steam locomotives could be heard off in the distance.

O X O X O

Sometimes at this point in the story my mother would laugh and say, "Now let me tell this part, John. This is my part." Then she'd tell how she and Katie relived all the fun they'd had at the Playmore. Margaret thought Johnny was close to male perfection. He was the first man she'd met who wasn't from Rock Springs, and she was fascinated with his soft

Scottish accent and his stories about Indiana. He seemed different than the other fellows she knew, more worldly, more sophisticated.

They dated for two years. Then in 1938, after asking Margaret's father for her hand in marriage, Johnny proposed.

She was only too eager to agree to begin married life with the man she loved.

A couple of weeks before their marriage, she quit her job at the UP Store to prepare for their wedding and the move to Winton. At that time, when a young woman married a coal miner, she didn't usually work outside the home for two reasons: miners prided themselves in being the sole supporter of their families and only a few jobs for women were available in coal camps.

Before he was married, Johnny spent very little money on himself. He saved money by having a $50 saving bond taken out of his check each pay period and sent a good portion of what was left back home to his parents. The first thing he did after Margaret accepted his proposal was to put his name on the company housing list. Then he and Margaret pooled their savings and paid $337 in cash for a brand new 1938 Ford Then they made a trip to Salt Lake City where they paid $800 cash at Sugarhouse Furniture for a house full of new furniture. They purchased their first kitchen coal stove from the renter who was vacating the house. The furniture was delivered to Winton by train and stored in a company warehouse until their house in Winton became available a few days later.

On Christmas Eve of that year, they were married in the Congregational Church in Rock Springs. Margaret was attired in a beautiful rose-colored dress with wine accessories and carried a bouquet of roses and sweet peas. Johnny wore a gray pin-striped suit and white shirt. They were attended by Bob and Jenny Wilson and married in the presence of close friends and her relatives.

Following the ceremony, they were guests of honor at a wedding dinner at the home of Margaret's oldest brother, John Copyak, and his wife, Mildred. After dinner, they spent their wedding night at Jenny and Bob's house and continued to live with them for a couple of days until they were finally able to move into their first company house.

○ x ○ x ○

Their first house was located on a hill behind Jenny and Bob's home. During the day, Jenny helped Margaret scrub floors and clean windows and cupboards, lining each with fresh shelf paper. In the evening, Bob helped Johnny paint, refinish the hardwood floors, and lay new linoleum in the kitchen. There was no indoor bathroom. Once they finished up inside, Johnny borrowed a company truck, and he and Bob hauled their new furniture from the Winton warehouse to their house. Then Margaret sewed draperies and curtains on the Montgomery Ward sewing machine she had bought when she was working at the UP store.

The front yard was nothing but rock, weeds, and dirt, and sloped down toward the street. Johnny again borrowed a company truck and hauled in stone from the surrounding hills. With the help of neighbors, he stacked one piece of flat sandstone atop another to create a rock retaining wall. Once that was done, he drove the company truck fifty miles north of Rock Springs to the town of Farson to dig up and load the rich topsoil he needed to plant a lawn. Finally around the newly leveled yard, he built a picket fence which he painted white.

To keep the Kentucky Bluegrass seed from blowing away in the strong Wyoming wind, he laid burlap bags over the entire yard and kept them wet until the grass began to grow. The company sent painters to put a fresh coat of paint on the outside of the house.

When Margaret's Grandfather Stephens in Rock Springs wanted trees removed from his property, Johnny dug up the trees to replant, hauling them one at a time back to Winton in the trunk of their car. Parts of the big trees hung out the back and dragged on the highway no matter how they tied them up. The first tree he transplanted was a weeping willow, similar to the trees in Indiana. He encircled the ground beneath each tree with brick and planted flowers in the ring. Sooner than they thought possible, the house began to look like a photograph from *Good Housekeeping* magazine.

Margaret said that as they worked she was impressed with his skills, his intelligence, and his sense of fairness and integrity. Johnny said he loved the way that Margaret appreciated even the smallest things he did for her.

Margaret and Johnny Nesbit,
before I was born, standing by
their home in Winton,
Wyoming, soon after they were
married.

From the beginning, they continued to send money from each pay-check to Johnny's parents before they paid even one bill.

○ × ○ × ○

People in the community socialized at the Winton Community Hall, the miners' family gathering place, where they watched movies on Friday and Saturday nights. Johnny and Margaret often went to Rock Springs to gather with her brothers and sisters at her parents' home. Sometimes in the evening, Johnny went to the camp pool hall where he shot pool with fellow miners and drank beer for ten cents a glass. Men in the pool hall talked about sports, fishing, local gossip, and always the mine.

About a year after they were married, Johnny drove to Cheyenne and took the certification test to become a journeyman electrician. Johnny had heard that the test took all afternoon and that plenty of men did not pass on their first try, so he'd taken it seriously and studied the manuals, but he lacked confidence since he'd never graduated from

My favorite photograph of Daddy and me, taken in the early 1940s.

high school. For him, the two weeks before the results were released moved like molasses. But one day there it was in the mailbox—his journeyman electrician certificate. He'd passed. It meant a considerable raise in pay and more responsibility.

Two years after they were married, on January 28, 1940, I was born. They named me Marilyn, after a little girl Johnny loved who'd come into the fruit market where he'd worked in Indiana. While Margaret was in the hospital, Johnny bought their first console radio and had it sitting in the living room when they brought me home. Two years later, my brother Jimmy was born, and two years after Jimmy, my brother Johnny arrived.

○ ✕ ○ ✕ ○

When I was not yet a year old, on December 7, 1941, the headlines in the Rock Springs *Daily Rocket-Miner* brought shock to everyone. Our country was at war. Racial bias, paranoia, and resentment ran throughout the mining communities. The UP had to take a stand.

On December 12, 1941, five days after the Japanese attack on Pearl

Harbor, the following announcement signed by Eugene McAuliff, president of the Union Pacific Coal Company, was posted for all employees of the UPCC:

The United States in now engaged in the greatest struggle that has taken place in history. The continent of Europe, much of Africa and Asia, and our own islands in the Pacific Ocean are aflame. . . .

The Union Pacific Coal company employs many men of diverse blood and religion. Whether we be citizens, native-born or naturalized or aliens who have not yet received citizenship (and some of our older employees cannot become citizens), we now have one common duty to perform— that of yielding unqualified allegiance to the Constitution and laws of the United States, and the fullest measure of respect and affection for the American flag. . . .

The notice went on to emphasize the importance of increased coal production for the war effort. Then it acknowledged the "assorted race and language" of miners and referred to the many national flags hanging in the Old Timers' Building. It concluded by clearly stating that no ethnic prejudice would be tolerated:

In our Old Timers' Building, there have been hung intermittently for years, all of the flags under which our people of assorted race and language were born. Many of these flags were but a short time since the symbols of a free people. They are the flags under which we of foreign birth first saw light. Today, many of these same flags have been torn from their place on this native soil by ruthless conquerors, but they yet remain the flags under which we were born, and as such, they will continue to appear from year to year. They, with us who are loyal to America, will stand second to the Stars and Stripes and flags of the nations which are joined to us in the fight against totalitarianism, the destruction of free government, and the enslavement of the soul.

What we wish to make clear is this—that each and every one of our employees must continue to give the fullest measure of obedience to our Constitution and our Government and respect for the American flag. Those who seek to do less cannot remain in our employ. Furthermore, there must be no rancor, hatred or prejudice shown by any one employee to another, who by accident of birth once lived under another government. All nations across the seas have in the past made definite contributions to the upbuilding of these United States. We today are all Americans, to work, to serve, and to pay as the necessities of our government may require. There is no other condition possible.

CHAPTER FOUR

ROOSEVELT SPEAKS TO MINERS

AFTER THE Japanese bombed Pearl Harbor, life in the United States, as everyone had known it, suddenly changed.

Right before the war, many men across the county were leaving the mining industry for other jobs. They no longer wanted to endure harsh and unsafe working conditions, substandard housing, religious and ethnic prejudices, labor disputes, violent strikes, layoffs, low wages, and the social stigma of being a miner.

Besides, mining was dangerous work. Newspapers reported devastating mine explosions where hundreds of men lost their lives in just one incident. These disasters left behind big families with no benefits and no place to live. Company housing was available only to employed miners and their families. After a fatal mine accident, the family of the deceased miner wasn't given much time to pack up and move out—the company needed the house for the next miner.

Mine jobs had no guarantees. When a field was exhausted, the mine was shut down and the miners were out of work.

As many men left mining and still more joined the army, suddenly the surplus of mine workers disappeared.

Immediately after Pearl Harbor, President Roosevelt asked the United Mine Workers and other unions to agree that for the duration of the war there be no strikes. Many unions signed no-strike pledges, but these were not legally binding, and the period from 1941 to 1945 witnessed more strikes than any other in American history, though many were short or wildcat strikes. Labor unrest was a serious concern during this time when the war effort needed to be uppermost in the minds of citizens.

John L. Lewis was the powerful leader of the United Mine Workers who ruled with an iron hand, and his name was a household word in the mining communities. He had threatened Congress that the mine workers would strike if miners didn't get the working conditions, pay, and benefits he asked for. His photo was often in the newspapers, easily recognized by his massive leonine head, shaggy eyebrows, firmly set jaw, and ever-present scowl. On the radio, his voice exuded power. He thrilled supporters and angered enemies. He once came to Rock Springs to address the coal miners of Sweetwater County and was given a western welcome.

Even the miners' children knew of John L. Lewis. The principal of Reliance High School, Ira J. Russell, told me of a teacher who asked her first grade class who could name the President of the United States. Many hands flew up and the teacher called upon one student. The first grader proudly answered, "John L. Lewis."

John L. Lewis threatened President Roosevelt that if conditions did not improve for coal miners, he would lead miners on a nationwide strike.

On May 2, 1943, when I was three years old, FDR made a broadcast to the American people, in particular to coal miners. Daddy said coal miners felt honored to be so personally addressed; it made their work seem important. Whenever the President was scheduled to make a broadcast, Daddy took our family into Rock Springs to my grandfather's house where we gathered with other relatives around the console radio in the living room to listen. Even though Grandpa was retired from the mines, he loved President Roosevelt and always listened to his broadcasts.

We kids sprawled out in front of the radio and colored in our coloring books, not making a sound. We could tell by the look on the adult faces how serious the presidential messages were. After Pearl Harbor was bombed, adults were often serious and children were fearful, especially of the Japanese and Germans.

The radio squawked and President Roosevelt began his address.

My Fellow Americans:

I am speaking tonight to the American People, and in particular to those of our citizens who are coal miners.

Tonight this country faces a serious crisis. We are engaged in a war on the successful outcome of which will depend the whole future of our country.

After detailing the accomplishments of the American farmers, soldiers, and production workers, he spoke directly to coal miners.

I want to make it clear that every American coal miner who has stopped mining coal—no matter how sincere his motives, no matter how legitimate he may believe his grievances to be—every idle miner directly and individually is obstructing our war effort. We have not yet won this war. We will win this war only as we produce and deliver our total American effort on the high seas and on the battle fronts. And that requires unrelenting, uninterrupted effort here on the home front.

A stopping of the coal supply, even for a short time, would involve a gamble with the lives of American soldiers and sailors and the future security of our whole people. It would involve an unwarranted, unnecessary and terribly dangerous gamble with our chances for victory.

Therefore, I say to all miners—and to all Americans everywhere, at home and abroad—the production of coal will not be stopped.

Then the president spoke of earlier reassurances by the four labor organizations that they would not go on strike during the war. A War Labor Board had been set up to settle disputes, but it had been unable to settle a recent one because the United Mine Workers refused to cooperate and had called a general strike.

The president continued and announced the unthinkable: the government had taken over the coal mines the previous day. Leon Lanoy, a miner who worked in the Stansbury mine, said that the miners were aware that the president had the power to do this when our country was at war but it was hard to imagine that it could actually happen.

President Roosevelt placed the blame clearly on the United Mine Workers and John L. Lewis. He said:

The responsibility for the crisis that we now face rests squarely on these national officers of the United Mine Workers, and not on the Government of the United States. The consequences of this arbitrary action threaten all of us, everywhere.

At ten o'clock, yesterday morning—Saturday—the Government took over the mines and called upon the miners to return to work for their Government. The Government needs their services just as surely as it needs the services of our soldiers, and sailors and marines — and the services of the millions who were turning out the munitions of war.

The war is going to go on. Coal will be mined no matter what any individual thinks about it. The operation of our factories, our power plants, our railroads will not be stopped. Our munitions must move to our troops.

Then the scratchy voice on the radio appealed to the patriotism of miners to return to work, regardless of their union's call for a strike. But he claimed to support unions and promised troops to protect miners and their families from any violence:

And so, under these circumstances, it is inconceivable that any patriotic miner can choose any course other than going back to work and mining coal.

The nation cannot afford violence of any kind at the coal mines or in coal towns. I have placed authority for the resumption of coal mining in the hands of a civilian, the Secretary of the Interior. If it becomes necessary to protect any miner who seeks patriotically to go back and work, then that miner must have and his family must have — and will have — complete and adequate protection. If it becomes necessary to have troops at the mine mouths or in coal towns for the protection of working miners and their families, those troops will be doing police duty for the sake of the nation as a whole, and particularly for the sake of the fighting men in the Army, the Navy and the Marines — your sons and mine — who are fighting our common enemies all over the world.

I understand the devotion of the coal miners to their union. I know of the sacrifices they have made to build it up. I believe now, as I have all my life, in the right of workers to join unions and to protect their unions. I want to make it absolutely clear that this Government is not going to do anything now to weaken those rights in the coal fields.

The first necessity is the resumption of coal mining. . . .

And he closed with these words:

Tomorrow the Stars and Stripes will fly over the coal mines, and I hope that every miner will be at work under that flag.

As Daddy told the story, the unrest ended with Roosevelt's speech, but I now know that on the national scene conflict between the United Mine Workers and mine ownership continued for years. Leon Lanoy, a former Stansbury miner, told me that he remembers that miners in the Rock Springs area continued to work under the government flag much as they always had, without experiencing picketing or violence.

Daddy, who as the foreman was a company man, did tell us how he hated the union because he felt it often hindered production.

He became disgruntled over designated job duties. Since he was a foreman, the union would not permit him to shovel coal because that was not the job of a company foreman. One older worker called Jonesy reminded Daddy of his own father. Jonesy couldn't work as fast as the younger miners and often lagged behind. Daddy sometimes picked up a shovel and worked alongside Jonesy to help him meet his production quota. Jonesy got tears in his eyes when Daddy helped him because he knew Daddy was helping him keep his job. However, if a union employee had seen Daddy doing this it would be reported as a union infraction.

Leon Lanoy related he didn't want union dues deducted from his paycheck and he never went to a union meeting, but he had to join to receive a paycheck. The union had a contract with the Union Pacific Coal Company, and miners could not work unless they joined the union.

But when Leon worked on his family farm in Oklahoma, he earned one dollar a day, working "ten to cain't" a day. He worked from dawn until dark. In 1946, he made fourteen dollars a day plus overtime working in the Wyoming coal mines and sent half his check home to his family. He needed the job so he was a union member. He jumped at the chance to "double over" (work one shift right after another), many times without food or drink for the second shift.

Sometimes miners missed a shift of work when the "mine pusher," who oversaw workers, saw an infraction and, after consulting with the union head, called a "wildcat strike"—an immediate strike right on the spot, whereupon all the miners had to leave the mine. This angered the miners who wanted to work.

Lanoy recalled a day when a company man was caught shoveling coal on the pan line. The pusher reported him, thus causing a wildcat strike. Sometimes a pusher didn't want to work because he had been drinking all night so he'd call a wildcat strike so he could have the day off. Others recall mysterious infractions around the holidays, which spawned short wildcat strikes resulting in more time at home for the holidays.

Each year in April, the union negotiated for higher wages, which often led to a short strike. During these times some children didn't get breakfast at home because of short supplies, but they usually brought sack lunches to school if only for appearance. At Reliance High School, the school started a hot lunch program, but most of the children still brought their lunches from home. The students clamored to carry their teachers' hot lunch trays back to the kitchen and along the way ate whatever was left on the tray. Teachers soon began to deliberately leave food on their trays for the children to eat.

During the war, ration stamps were issued for sugar, shoes, and gasoline. Miners who lived any distance from the mine had to ride together to conserve gas. People with trucks got more gasoline ration stamps and shared them with other miners, even though the practice was illegal. Larger families received more stamps and, if they had extra, they offered them to teachers and others. Nylon hosiery was impossible to buy because nylon was needed for parachutes, and bedsheets were in short supply. Automobiles had to be ordered. Daddy ordered a Hudson and drove to Detroit to pick it up to save money. Single miners valued lunchmeat for lunches, because it gave them a break from their routine crackers and cheese, which didn't provide much energy for a hard day's work.

○ ✕ ○ ✕ ○

The President's call was answered by former miners all across the nation. Thus, Rock Springs with its rich coal reserves attracted another influx of miners to fill jobs with Union Pacific Coal Company.

Since Daddy was a company man, he went to work each day whether the mine was on strike or not. The short strikes caused little violence, and we children didn't notice any disruption in our routine child's play or our friendships.

But even at a young age, we children were aware of what was happening at the mine. If there was a mine accident and the whistle blew, we ran to the mine right along with every adult in camp.

CHAPTER FIVE
COAL CAMPS THRIVE

HEEDING THE words of the President, disheartened miners from all over the country who had stopped mining coal, cast aside their grievances and returned to the mining of coal.

Though labor struggles continued nationally, the Wyoming coal camps were isolated from them by distance and slow communication. A big influx of miners from Europe, Oklahoma, and Arkansas arrived in Rock Springs in search of any kind of work and housing and joined the melting pot of miners already there.

Union Pacific enthusiastically hired miners from many foreign countries. The cultural and linguistic differences among the immigrant workers made the formation of labor unions more difficult, and the company could better control employees. At this time, Rock Springs was listed as the third most ethnically diverse town in the nation with fifty-six nationalities, behind only San Francisco and New York.

In the beginning, the miners who came to the area were single men or men who had temporarily left their families behind. They lived in boarding houses or with new friends who had extra space in their homes until they got enough money ahead to send for their families. This forced arrangement of being thrown together helped the miners assimilate.

Miners not only saved money for their immediate families, but also faithfully sent money home to parents. Workers were motivated to live on very little money.

○ × ○ × ○

Many newcomers had no idea what to expect in Wyoming and upon arrival felt like they had reached the end of the earth. Their letters almost always mentioned the absence of trees and the presence of miles of sagebrush in every direction.

Some miners also confronted social isolation. Many could not speak English. Because of the war, miners and their families who were thought to be of Asian or German heritage faced intense prejudice and segregation.

Some of the new laborers had been miners in their homelands, but most, like the Slovenians had been farmers or urban dwellers. Slovenian immigrants had to make the transition from a society where manpower, not steam power, dominated industry. U.S. mines were increasingly dependent on machines, and the immigrants had to adjust to new mining techniques which reflected America's growing industrial strength.

Formal training was minimal. Miners learned from those men working beside them. Accidents were all too frequently part of the learning experience. Immigrants were more likely than native miners to face injury because their bosses used a language they didn't understand. Bosses traditionally tested new miners with backbreaking and demeaning jobs to see how much they "wanted to work" and used American profanity which foreigners didn't comprehend. The bosses, with their own ethnic prejudices, came down harder on certain minorities than others.

Company housing was often scarce and rundown. Much was officially classified by the United States government as unfit for habitation. Immediately after signing on for a job, new miners stood for hours in lines to add their names to a housing list. Most housing was nothing more than a "tar-paper shack" quickly thrown up on unlit, dirt streets.

Yet miners, eager for good paying work and sometimes unaware of the challenges of the climate and conditions, kept coming.

UP worked to remedy this situation. Almost overnight, they developed housing from old boxcars and set them just outside Rock Springs in the White Mountain Addition. Twelve-by-forty foot boxcars, with the running gear removed, were plumbed, wired, and placed on cement foundations, two on each slab. Each boxcar housed one family. No insulation was added to the thin walls designed to enclose freight or livestock, not people. The coal stove in the kitchen provided heat for both cooking and warming the entire "house." UP delivered coal and dumped it on the ground in an unsightly pile outside each unit.

These boxcar dwellings were also sometimes moved onto a piece of rail track which ran parallel to the regular coal train track leading into Reliance. Their running gear remained underneath, and they were not set on slabs, which made it difficult to climb up the steps to get into the dwellings. Unlike the boxcars in Rock Springs, they did not have indoor toilets so the residents had to use outhouses.

Vice President Dick Cheney's acceptance speech at the 2004 Republican Convention mentioned that his grandfather, who had worked as a cook for the Union Pacific Railroad, had at one time lived in one of these boxcar dwellings.

My childhood friend Geraldine Guigli moved to Rock Springs from Walsenburg, Colorado, with her mother and stepfather, Alice and Videsto Palazari, for his new mine job. A boxcar was the only place available to live. She talked about the horrible living conditions and how scared she was to ride the school bus to Rock Springs for grade school. After school one day she missed the bus and was terrified sitting on the school porch waiting for her parents to come and get her. Since her father took the car to work each day, she had to wait until after five for him to get off work. She felt lost attending school in the big city of Rock Springs. She often cried and begged her parents to move back to Colorado.

As primitive as the boxcars were, families scrambled to be assigned one just to have a place to stay, but the housing was never comfortable. Geraldine described how during the winter, they slept in their winter clothes under heavy wool blankets and still felt like they were freezing. The glass panes in the few windows were often thick with frost because insulation was non-existent or crudely homemade. Each morning her mother scraped hard frost off the inside windows even though the coal stove was fired up.

In the summer, the boxcars were so hot that Geraldine could hardly remember ever being cold. When it rained the entire development was a sea of mud. Miners' wives had a full-time job just keeping the boxcars at a reasonable temperature and fairly clean.

The floor plan of the boxcars with bathrooms consisted of a small kitchen-living room in the middle, with a bathroom on one end, equipped with a small coal stove to heat water, and a bedroom at the far end. Linoleum covered the floor boards throughout. Geraldine described living in this housing as "a nightmare."

Families with babies placed the baby bed in the kitchen near the stove and tightly wrapped the little ones almost all the time. Most mothers nursed their infants. When that wasn't possible, they made their own formula using one can of canned milk, twenty-eight ounces of boiled water, and three tablespoons of dark Karo Syrup because it contained dextrose maltose and was thought to be easier for babies to digest. When babies were brought home from the hospital, they wore a cotton bellyband around their tummies until the umbilical cord healed. Those same bellybands were washed and used time and time again whenever children had colds. Mothers rubbed Mentholatum or Vicks ointment on their children's throats and carefully wrapped the bellybands around their necks to hold the warmth of the salve.

Families congregated in the kitchen around the stove. In the summer, bathing was done in the bathroom, but in the winter, bathing was done once a week, usually on Friday night, in a round galvanized tub set in the middle of the kitchen floor because the bathroom was near freezing. The wife boiled water in kettles on the stovetop and poured it into the tub then cooled it by adding kettles of cold water from the kitchen sink. Everyone took a turn, using the same bath water. The youngest children were bathed first. Hot water was added sparingly. The man of the family was able to shower in the mine bathhouse, thus only the women and children bathed at home.

Geraldine recalled that the boxcar families did everything they could, with whatever they could find, to make a nice home. Sticks of dynamite, used for blasting in the mine, came packed in sturdy wooden boxes. The miners packed home the boxes to use for cabinets by stacking one on top of another. When families fabricated better furniture they did not have any trouble finding a neighbor who was glad to have their hand-me-downs.

Geraldine's mother and many other women braided rugs from rags to decorate their homes or to sell to make extra money. Very few families had bedspreads, so they prized Army blankets and handmade quilts. It was a treat for a housewife when she could afford a new piece of oilcloth to cover the boards on the kitchen table because oil cloth "wiped down so nice." Many wives hung curtains made of muslin for a little privacy. Most household items could be purchased at the company stores, including cast iron skillets and pots for cooking—if you had the money. The company store in Rock Springs had a larger selection than the camp stores, so wives looked forward to trips there.

Whenever our family went into Rock Springs, the road took us by the White Mountain Addition. I would look out the window of our car in amazement at the ugly boxcars. I felt sorry for all the little children I saw playing in the dirt yards and living in those crowded boxcars.

○ × ○ × ○

Men outnumbered women in Rock Springs and houses of prostitution thrived. Ladies of the evening or "K Street ladies," as they were called, lived in apartments in old buildings, often above local businesses. A red light earmarked a street-level entry door with a flight of steps leading upstairs. The establishments were listed in the telephone directory under "Furnished Rooms for Rent," with names such as the Lux Rooms, the Capitol, or the M&M Rooms. On Saturday night women sat on their window sills with red velvet drapery blowing in the air around them, walked the main streets, or occupied stools in saloons. Many were beautiful women, some very young, and they wore low-cut, fancy dresses, heavy make-up, fishnet hose, spiked heels, and lots of perfume. When they shopped in the local stores they did not meet the eyes of the wives of the community who would never have smiled or spoken to them.

All of us kids knew about the women, but the first time I saw one I was about ten years old. My brothers and I peeped into the window of the wild Southpass Bar. Momma always told us never to linger when we walked past the bar with its loud music and frequent fights which sometimes spilled right out onto the sidewalk. But this time I looked in and saw a beautiful woman sitting on top of the bar, scantily dressed in a glamorous red dress with feathers. She looked like a movie star sitting with her legs crossed, wearing fishnet hose and red spike heels with ankle straps. I was fascinated.

On the ride home, I told my parents that when I grew up and made my own money I was going to buy a dress just like the lady in the Southpass Bar. Daddy gasped and firmly told me to get that crazy idea right out of my head.

The ladies of the night often stood on the street corners on Saturday nights and were easily recognized by their dress. If I pointed them out to Momma, she would say, "Those damn hussies."

One day my mother and I went into Union Mercantile and had the shock of our lives. A prostitute stood with a grungy older man looking into a lingerie display case. Suddenly, she stepped behind the counter, knelt down and proceeded to remove her blouse and bra and try on one of the bras from the display case. She did it so fast without any inhibitions that Momma and I just stood with our mouths hanging open. The clerk waiting on her was as shocked as we were, but didn't say a word. Momma said prostitutes spent a lot of money on clothing in that store, so the clerk probably didn't want to cause a fuss.

Each mining camp had one saloon which was owned by the UP and leased to the operator. But in Rock Springs each block on the main streets had at least two privately owned saloons. In the 1940s and '50s, miners could buy hard liquor or a ten-cent glass of cold beer, shoot pool, sit in on a game of poker, sometimes with high stakes, talk to friends, or find a warm, welcoming woman.

○ x ○ x ○

Then almost overnight news traveled throughout the mining camps that UP was about to embark on something big. The UP planned the development of a new mine with a new community.

The wives spread the news with great anticipation. And the most exciting part to the women was that the company would construct approximately 110 four- and five-room, wood-framed houses, set on foundations with indoor bathrooms and full basements. It would be a model community and a wonderful place to raise children.

CHAPTER SIX
STANSBURY LEAPS TO LIFE

As soon as Daddy heard about the new mine—Stansbury—he knew he had to make a move.

Mines closed either because of changing economic conditions or because they ran out of coal. Regardless of how thick a coal seam was, it eventually "pinched" and became too thin to mine. The closure of a mine sometimes resulted in the abandonment of an entire town. Daddy suspected that before long Winton would become one of those towns. He brought home all the information UP made available about the proposed new mine and explained some of the details to Momma, both about the mine and the housing. Ever since the war had broken out, UP had been talking about the development of a new large-capacity plant to augment the declining production from the old mines in the Rock Springs and Winton districts. With government funds, Stansbury was to become that new operation. Daddy knew it would be smart to ask for a transfer. During the start-up phase, practically every weekend he and Momma drove to the Stansbury site to observe the sinking of the new mine and the construction of the new homes. Momma was beside herself just imagining what it would be like to live in one of the brand new, modern houses with an indoor bathroom.

Stansbury was rumored to be one of UP's model communities, except with one big difference; the new housing would be constructed before the mine went into full operation. This new mine would employ approximately a thousand miners working three eight-hour shifts, seven days a week.

Stansbury would derive its power from the Rock Springs power plant. Water would come from the neighboring town of Reliance, which had a well of an unusually large flow and clear, potable characteristics.

During the sinking of the mine and construction of the camp, the Rock Springs newspaper reported the progress. Miners flocked to get their names on the list to work at what was predicted to become one of the most productive bituminous coal mines in the country. Stansbury coal was considered a reliable high heat, low ash, steam coal for steam locomotives. Most Wyoming coal is sub-bituminous, but this was atypical as it had low moisture levels and hence a much higher BTU. The Stansbury coal was Bituminous Rank C—good stuff.

Miners' wives, excited at the prospect of spacious new houses with indoor bathrooms, pushed their husbands to sign up to work at the new mine. Their current housing suddenly seemed worse than ever.

Two months before the mine opened, UP officials asked Daddy to be the underground mine foreman at the Stansbury mine. He quickly accepted the offer, and Momma couldn't wait to move. One day he took her to see some of the finished houses. He told her she could choose the house she wanted and which street she wanted to live on. Momma was ecstatic. Years later she would still get tears in her eyes as she remembered the day she walked through one of the new houses. She said she couldn't stop babbling in joy: "Oh, my God, John! I have waited my whole life for a house like this! This is a dream come true. Just look at the beautiful hardwood floors! And, a formal dining room for the holidays! Oh, John, when can we move?"

Daddy wrapped his arms around her and told her to start packing as soon as they got home. Momma chose a five-room house on a street that led to the future elementary school. She said the day she and daddy hauled the first load of their possessions to the Stansbury house was one of the happiest days of her life.

The modern, white-framed houses were spacious and came in two different styles. The four-room homes had two bedrooms, kitchen, living room, and bathroom with a detached garage and were available to hourly employees.

The five-room houses were similar but somewhat larger and had a formal dining room, an attached garage, and a big front porch. These

The new company town of Stansbury featured four-room houses for hourly employees and five-room houses for salaried employees. The houses with the entry and steps in front are four-room houses. The house in the middle is a five-room house with the entry out of view, near the attached garage. (Department of Interior, Russell Lee)

five-room units were for salaried employees: mine foreman, unit foremen, and engineers. The biggest house in the camp was that of the mine superintendent. Since Daddy was the mine foreman, we were assigned one of the five-room units.

All the houses had hardwood floors, basements, and most importantly, inside bathrooms. To this point, houses in every other mining community had only outhouses. The men showered at the mine bathhouse. Families in houses with no indoor bathrooms bathed in a way similar to the boxcar dwellers—in large, round galvanized tubs brought in from laundry rooms or back porches and set in the middle of the kitchen floor. If small children were bathed, the tub was often set up on two kitchen chairs so parents could easily reach them. After baths, the tub was emptied using buckets to scoop out the water, which was poured into the kitchen sink. Then the tub was wiped clean and returned to the laundry room where it hung on the wall until needed for laundry.

I was three years old when we moved, Jimmy was a year old, and Johnny was born after we moved to Stansbury. Momma took a picture of me pushing my doll buggy around in the dirt yard in front of the new house. She told me that my favorite pastime quickly became standing on our front porch and jabbering incessantly to all the miners coming home from work each day.

Within a short period of time, many miners started work, and families moved into the new houses. Soon the sound of steam locomotives transporting coal continuously day and night could be heard throughout the camp. The miners' children loved to watch as the trains entered or left camp with black smoke billowing from the stacks. We also loved to hear the train whistle, and many times ran to the side of the tracks pumping one arm up and down as a signal to the engineer to keep blowing the whistle. And he would.

Stansbury wasn't your typical Wyoming mining town of the 1940s. It was indeed UP's model community and everything in it was new.

A railroad car is being loaded at Stansbury mine tipple in 1953. The multiple sets of tracks can be seen approaching the loading facility. Built of steel, the tipple had a loading capacity of 500 tons an hour. (Rock Springs Rocket-Miner)

Union Pacific, with financial help from the government, constructed not only the houses, but a general store, community hall, post office, doctors' clinic, boarding house, and a brick schoolhouse. The mine had first class facilities such as offices, a large bathhouse, a shop, and a tipple.

The focus of mining was the underground operation, but the purpose of the mine was to ship coal back East. The tipple was a building positioned down from the mine opening, built high above the ground, extending out over the railroad tracks. Here the underground coal cars, each with a capacity of five tons, dumped or "tipped" into a metal pan that carried the coal to a "picking table" where the coal was sorted according to size. Old men, young boys, and sometimes women stood over the picking tables separating the coal from slate or stone and throwing out any material that would not burn. Years later, noisy, steam-powered shakers were used.

Once sorted, the coal was loaded through chutes into waiting railroad coal cars. The process of moving coal inside the tipple was accomplished principally by gravity and back-breaking effort. The moving coal created clouds of black dust that covered everything and everyone. This surface operation was a difficult and relatively elaborate job.

○ ✕ ○ ✕ ○

Momma really appreciated the indoor bathroom and told us children often how lucky we were to have it. She remembered what it was like being raised in a Rock Springs company house without indoor plumbing. Since indoor bathrooms weren't usually available then, UP made special arrangements for the miners' families. Most notably the company reserved days for miners' families to shower in the bathhouse. Momma said she and her sisters never forgot the thrill of taking their first shower. She said taking a shower was something many of the immigrants had never done before either, and they had to get used to it. She recalled those special times when she and her brothers and sisters followed their mother down the street leading to the bathhouse carrying their towels and clean clothes to change into.

When I was a girl, we children were never allowed to take showers in the bathhouse as we had houses with plumbing. However, occasionally on hot summer days, the bathhouse manager at Stansbury allowed the camp kids to play under the shower heads in the big shower room at the bathhouse to cool down. We were instructed to bring our own towels, and we ran and splashed while the miners weren't there. The only other time the camp kids showered was when we used the swimming pool at Rock Springs High School.

UP rented the new houses to the miners for $40 per month, a rate that was never raised. The rent was deducted from the miners' paychecks each month. That $40 included sewer and water. The only things miners had to pay for separately were electricity and telephones. Our phone number on the five-party line was 04J4.

Coal and electricity fueled the houses, with coal used in furnaces, hot water heaters, and cook stoves. Each house had what we called a

*Johnny Nesbit, my father,
with us kids: me, John Jr.,
and Jimmy.*

coal bin in one corner of the basement, with a covered chute that led outside for coal deliveries. To the annoyance of our mothers, we kids gathered to watch on the days the one-ton company dump truck delivered the coal to nearby houses. Once a month the truck arrived and the driver hand-shoveled the coal into the chute of each house. Mother stewed about the thin film of coal dust that infiltrated the entire upstairs of the house on the days that coal was delivered. She remedied the situation by hosing down the inside of the coal bin in the basement with water right before the coal was delivered. Her idea worked so well to suppress dust that soon every housewife in camp did the same thing. Residents ordered coal deliveries at the local mine office at $20 a ton, deducted from the miner's paycheck.

This coal burned very hot. On a regular basis someone had to clean the ashes out of the furnace and stoves. Daddy did this most of the time, but sometimes Momma had to do it. She said she could feel the heat deep in her chest as soon as she began shoveling the ashes out of the furnace. She put the ashes into a coal bucket, carried it up the basement stairs to the outside, and dumped it in the garbage can in the alley behind the house. The smoldering ashes burned any garbage in the can. The UP provided one garbage can for each house. These cans were placed at the far edge of the backyard, so they did not create a hazard when they caught fire. Once a week, a company dump truck drove through the camp, picked up the burned garbage, and hauled it to the camp dump on the outskirts of town. The charge for this service was part of the monthly rent.

In the basement, a four- or five-foot concrete retaining wall ran the length of the back wall. To make use of the space above the retaining wall a short, timbered wall was installed, creating a long "cubbyhole" with a wooden access door at about eye height. The floor of the cubbyhole was dirt. Kids found this space fascinating and scary.

One could not stand up in the cubbyhole as it was just high enough to crawl around in. One light bulb hung in the middle, which was turned off and on by a switch to the side of the latched, wooden access door which was the only way in or out. Daddy laid wooden planks, that he got from the mine, across the dirt floor so no one had to crawl around in the dirt. Momma stored her empty canning jars and Christmas decorations there, and Daddy stacked kindling for the furnace in empty dynamite boxes from the mine and stored them there as well. We kids heard that snakes crawled into the cubbyhole from underneath the foundation of the house, and Momma hated the very thought of that. When she needed anything from the cubbyhole, she had Daddy get it for her. My brothers and I never went in there after we saw Daddy bring out a bullsnake one day.

The coal furnace was in the center of the basement and beside it was a small "monkey stove" which was used to heat hot water. The wives did laundry in the basement using a conventional wringer washing machine, wooden scrub boards, and two galvanized wash tubs for rinsing set on a portable stand beneath the water faucets.

Anyone who visited a coal camp on a Monday saw laundry hanging on clotheslines behind each house. The women spent most of every Monday completing this backbreaking chore, which began right after the men went to work. Momma began Mondays by gathering, preparing, and then refrigerating the makings of a slow-cooked evening meal, oftentimes homemade stew, chicken and noodles, lima beans and ham, or beef soup. Later she put the ingredients into a cast iron dutch oven to simmer slowly on the stovetop all afternoon while she did laundry.

She went to the basement and lit some kindling in the monkey stove. When the kindling was burning well, she added coal to heat the hot water heater so she would have plenty of scalding hot water for laundry. She kept this fire going the whole time she was doing the wash. The way she did it was really an art. To save fuel, she always tried to conserve hot water by changing the hot, soapy water in the washer only when it started to get dirty. Then she drained all the dirty water with a hose connected to the washing machine into a drain in the basement floor.

Her first load of wash was sheets, white clothes, underwear, and socks, which she pre-scrubbed on the wooden scrub board. Since the first load had Clorox in the water, she changed the water before she washed the colored clothes.

She liked to add liquid bluing to the rinse water for special items—like men's white dress shirts, women's blouses, and pillowcases—so they came out snowy white. Momma hand-rinsed these items in warm starch water, which she had prepared in a big pan on the monkey stove.

The last load was mine clothes, often referred to as "pit" clothes, which Daddy brought home to be laundered every Friday after work. His Levi bib overalls, shirts, and jackets were usually filthy, covered with coal dust, grease, and sweat. Momma soaked them in rock salt the night before and then laundered them in White King laundry detergent which had a reputation for really getting the grease out of work clothes. Finally, she hung the freshly laundered clothes on clotheslines outside—unless it was freezing cold, then she hung things in the basement on the clotheslines Daddy had made for her.

Before a winter storm or when the Wyoming sandstorms came up, Momma ran to snatch the clothes from the clotheslines and hung them in the basement or over wooden chairs near the kitchen stove. When I helped her bring in the clothes, I loved to stick my face into a basket of sheets fresh off the line and inhale the fragrance. To this day I've found no cleaner smell than that of sheets dried in the fresh outdoor air.

After the clothes were dried, Momma "dampened" the starched clothes for ironing, a process unique to that era. She sprinkled each item with cold water from a pop bottle topped with a cork-based metal cap with holes in it. Then she carefully rolled each item tight and placed them in a zippered plastic bag to be ironed the next day. If she ironed them before that, they would be too wet. If she waited too long, the wet clothes would mildew, and she'd have to start the process all over again, beginning with washing.

I loved to watch and help Momma on wash days when there was no school. I especially liked putting items through the wringer on the washing machine, heeding Momma's words to be very careful not to get my fingers caught in the wringer. I wanted to do anything to make her workday easier. On those days, it was my job to scrub all the white socks on the wooden scrub board before putting them in the sudsy washing machine. She showed me how to mix bluing and starch water for Daddy's white dress shirts which he had to wear to mine meetings in Rock Springs. When I helped her hang the laundry outside, I had to be sure to hang everything in order, not just any old which way: socks, pillowcases, sheets, shirts. Sometimes she let me iron pillowcases and handkerchiefs, and I begged her to teach me to iron Daddy's shirts. Finally she did, and I got good at it. I even did ironing for my teacher and his wife, the Lanoys, to make spending money.

Momma reserved a separate day for washing blankets, bedspreads, rugs, and sheer white lace curtains, used to dress the windows of the miners' homes. These curtains had to be starched in a heavier starch and stretched on big wooden hangers. I remember how happy Momma was the day Daddy surprised her with her very own wooden curtain stretcher, a large wood framed hanger the size of a single panel of curtain. It sat on four legs and had small nails about an inch apart all along the four edges. After these lace curtains were stretched and back on the windows, Momma's hands would be pin-pricked, red, and sometimes swollen and sore. But she and her friends agreed that freshly starched and stretched curtains looked beautiful in the windows. Momma wouldn't let me help her do this until I was in my teens.

The miners' wives prided themselves on their canning. Most of the camp women stored their homemade canned goods, as well as food purchased at the grocery store, on storage shelves under the basement stairs. Momma made a colorful display of canned fruit and vegetables all lined up in sparkling Mason jars along the shelves in our pantry.

One of the chores Daddy did was to fill the coal bucket, which sat by the kitchen stove, each night before we went to bed so that there'd be plenty of coal to fire-up the kitchen stove in the morning.

Miners could buy anything they needed at the UP store located in each camp or at the main store in Rock Springs, where they had sales on the weekends. Miners could charge all their purchases at these stores and set up monthly payments, deducted from their paychecks. Thus, the UP controlled yet another facet of the miners' lives. When some miners picked up their checks, the dollar amount showed as "railroad tracks," two lines like an equal sign, with no numerals. That meant they received no income for that pay period because they owed the company store more than they had earned.

Edmond Jefferies, who managed the big UP store in Rock Springs, had empathy towards the miners and advanced broke miners $20 cash, which was added onto their bill to be deducted from the next payroll check. When a man worked for the UP, it was easy to accumulate more and more debt until, as the Tennessee Ernie Ford song recounts, he truly "owed his soul to the company store."

What the Stansbury UP Store didn't stock was probably available at the Rock Springs UP Store which had more variety: shoes, clothing, groceries, freshly cut meats, bakery items, gasoline, sewing materials, hardware items, furniture, bicycles. Children dreamed of the day they would receive their first brand-new bicycle from the store. Many Stansbury families seldom went to Rock Springs to shop because they enjoyed the convenience of getting everything they needed right at their local UP store. The Rock Springs store made daily deliveries out to the camp stores, so people could request special items which were delivered the next day.

Miners knew what it was like not to have any money, and they didn't like the feeling. So, in an attempt to have some type of savings plan, many had a twenty-five or fifty-dollar bond taken out of their

The Cummings Boarding House provided excellent food for hungry min-
ers. Ann and Jim Cummings stand in front. (Coal Camp Reunion
Handbook and Sweetwater County Museum)

check once a month. Daddy always had a fifty-dollar bond taken out.
The reason was two-fold: to save something for a rainy day and to help
the country by buying government savings bonds. Miners were paid
every two weeks, and men lined up at the mine office after work on Friday, picking up their checks to take home with pride to their wives.

Stansbury quickly became a thriving community with most activities
and services held in facilities owned by UP. Teachers received good salaries
from the Reliance School District #7, which attracted many young edu-
cators right out of college. Unlike many of the older mining communities,
the Stansbury school had brand-new books, teaching equipment, and a
playground with a full set of swings, a merry-go-round, and slide.

○ ✕ ◇ ✕ ○

Many single miners boarded at the Stansbury Boarding House, operated
by Jim and Ann Cummings. I loved to stop by the boarding house in
the evening after supper to visit with Ann and Jim on their front porch.
Sometimes I'd get lucky, and they'd treat me to a piece of Ann's famous
homemade *potica* (sweetbread nut roll) dessert.

Miners could always find a hearty supper at the boarding house and felt lucky if Ann served her famous potica. (Sweetwater County Museum)

Their daughter, Jo Ann Cummings, remembers that Jim and Ann began leasing and managing the boarding house in 1945. Ann's mother, Kata Krpan, had operated a boarding house in Reliance, so Ann's knowledge of running a boarding house came naturally. Most importantly she was an excellent cook who put hearty meals on the table. Jim worked at the mine in addition to helping Ann at the boarding house.

They put in long hours to accommodate around fifty boarders, men of all ages from all over the world. Many lived there until they saved enough money to get a place of their own and send for their families. Some were seasoned miners; others didn't have a clue what they were in for, but boarding house living helped new miners assimilate.

Because the camp was booming, single miners also stayed in two houses across the street from the boarding house and came over for meals and laundry. At that time, Ann prepared meals for about a hundred miners and carpenters every day. She cooked two breakfasts: one for the miners going to work and one for the men coming home after a night shift at the mine. She also prepared lunches for the men, washed lunch buckets, and packed them up.

Ann served four meals and packed two sets of lunch buckets every day. Even as a child I marveled at how Ann and Jim's two children, Billy and Joanne, helped out by peeling one hundred pounds of potatoes each day, cleaning the dining room, and doing the dishes. Jim and Ann served all meals family-style, were never late with a meal, had plenty of food, and welcomed anyone who happened by.

Ann and Jim organized potluck dinners, cakewalks, and handed out the best treats at Halloween. Mining families knew they'd lend a helping hand.

Jim and Ann worked extremely hard themselves and expected nothing less of their employees. Some said they were difficult to work for because they expected a lot from employees, but most agreed that they learned a work ethic there that served them well all their lives.

○ × ○ × ○

The preliminary development of the town was completed in January 1944, marking the first milestone in the 5,000 ton daily capacity underground mining operation. Our first year in Stansbury, residents endured one of Wyoming's severest winters. The freezing wind blew right through the pretty wool coat I got for Christmas.

The Stansbury Community Hall opened its doors to the public for the first time March 16, 1946, with a big community dance. The Cummings family provided the sandwiches and pop. All the buildings in town were owned by UP and leased out to the operators who were often mine families. The wives ran the businesses with help from their families. Each of the surrounding mining communities also had a general store, medical clinic, community hall, post office, saloon, and pool hall, but Stansbury also had a bowling alley, soda fountain, and barbershop, and all the buildings were brand new.

The community hall was next to the UP store and was the gathering place for all camp social events. It was open only for events scheduled through the mine office; otherwise it was locked up. On Saturday nights families gathered in the big upstairs room for movies shown at no charge and for dances with live local bands. The various clubs and organizations, such as the Homemakers' Clubs, 4-H, Girl Scouts, and Boy Scouts, met in the hall, and the men gathered here for their weekly mine safety meetings. During the Christmas holidays, the school held its annual Christmas program in the upstairs room, using a stage constructed solely for this purpose.

Emma and Louis Tomassi managed the one-lane bowling alley, barbershop, soda fountain, and bar in the downstairs of the community hall. Louis also worked in the mine.

Like many of the camp kids, my first savings plan involved buying ten-cent savings stamps at the post office and pasting them in a special booklet. When filled, the book had a value of $18 and could be taken to any bank and either cashed out or used to purchase a savings bond. Enid Matson, the Stansbury postmaster, welcomed us when we took our dimes in to invest. Her husband, Raino, worked in the mine office.

The thought of the Stansbury store makes my mouth water to this day. Louis Milojevick and his wife Betty kept the store stocked with items that adults needed, but also stocked things children coveted. Betty had grown up in Winton as Betty Wariner so she remembered what it was like to be a camp kid with only a few pennies to spend. The polished glass display cases were filled with a bewildering assortment of sweets: chocolate licorice, Monkey Bars, Idaho Spuds, Sen Sen, and Double Bubble Gum. We kids loved wax whistles and wax bottles filled with sweet syrup, Sugar Straws, Nibs, Dots, Black Crows, Big Hunk, Hershey bars, Walnettos—and all kinds of peanuts and gum, including the popular Beaman and Black Jack gum. I bought my very first bottle of Halo Shampoo at the Stansbury store. I loved the clean fragrance of that shampoo so much I continued using it until they quit making it, some years after I graduated from high school.

○ × ○ × ○

The medical clinic, in a small house across from the boarding house, opened every afternoon after three o'clock when company doctors drove out from Rock Springs. Dr. Kos, Dr. McCrann, Dr. Bertoncelj, Dr. Bergoon, and Dr. Muir each made the trip at various times. UPCC subsidized the service and the prescriptions. Many times we got our prescriptions directly from the doctor.

During the 1940s, the doctors made house calls when we children had childhood diseases like measles and mumps. Dr. Muir was our favorite because he even took time to read stories to me and my brothers.

Every year the UP paid for the testing for tuberculosis. A big semi-truck and trailer containing a huge x-ray machine arrived at each mining community and parked in front of the store. Adults lined up to have a chest x-ray for detection of this common disease of miners.

At the beginning of each summer, mine doctors inoculated children against tick fever. Everyone dreaded those shots, but our parents made us have them because the sagebrush in the surrounding hills was loaded with ticks. A rumor spread among the kids which scared us even more than the shots. As the story went, while giving one of these injections, a doctor had broken a needle off in someone's arm. We envisioned this grisly possibility and discussed it endlessly. This only heightened our fear of getting the shots.

The company also paid for yearly dental fluoride treatments usually at Reliance High School under the direction of dentists from Rock Springs. We children received a cupful of liquid and were told to swish it across our teeth. UP also paid for inoculations against childhood diseases including, later, the dreaded polio.

○ × ○ × ○

Several traveling salesmen had routes that took them to the homes in the mining camps. Momma and the other wives particularly loved the spices and teas the Jewel Tea salesman sold. Since Daddy was a Scotsman, we drank a lot of tea at our house. Jewel Tea gave points for purchases, and, with the points, housewives could acquire Jewel Tea dishes, which today are collectibles.

One summer, June Haver, a beautiful Hollywood movie actress, came to Stansbury. Everyone in camp gathered at the UP Store to get a glimpse of her. She was engaged to marry Dr. John Duzik from Rock Springs. He brought her out to go down inside the Stansbury Mine. No one could believe that the company was actually going to let a woman go down into the mine for it was a long-held belief that a woman in the mine brought bad luck. But they let her go down.

When I heard a movie star was coming to our town, I was ecstatic. My friends and I stood in front of the UP store, along with everyone else in camp, holding our autograph books and waiting to get a glimpse of her. All of a sudden a fancy car drove up, the car door opened, and there she was. We all ran up to her starry-eyed, and she signed each and every one of our autograph books

She was a petite young woman, much thinner that she appeared in films, resembling a beautiful doll in a tissue-lined box standing alongside Dr. Duzik who was a Rock Springs hometown fellow. Since Miss Haver didn't have the proper work clothing to go underground in the mine, the Stansbury UP Store let her borrow what she needed. When she came out of the mine, she returned the clothing to the store. It was a perfect fit for Gloria Fabiny who worked at the store so she bought the things as keepsakes. Everyone enjoyed visiting with this down-to-earth celebrity, and we were saddened to learn that Dr. Duzik died shortly before they were to be married. Miss Haver later left acting to become a nun for a few months before she married actor Fred McMurray.

o x o x o

When I was five years old, our family took our one and only vacation back to Indiana so that Daddy could fulfill a promise he made to his parents—to bring running water into their house.

Daddy cried when he saw his parents sitting on the front porch waiting for us to arrive. My brother Jimmy and I, anxious to finally get out of the confinement of our car, ran through the big willow trees in the yard until we discovered an old tree house Daddy had built when he was a young boy. Once we found this, we spent every waking moment playing in it along with the cousins we met for the first time. Every time I saw Daddy he was smiling. He'd often stop to put his arms around his mother.

Grandmother Nesbit had special gifts waiting for my brothers and me: a big box of cars made out of rubber for my brothers and a white purse with a gold chain for me. Each night before I went to bed, I carefully hung the purse on the post of my bed.

Daddy and Momma had saved for months to buy the needed plumbing supplies. With the help of his brothers and friends, Daddy worked hard all day and into the night for the entire two weeks to complete the project. They plumbed the house and put in a bathroom and a sink in the kitchen. His mother cried with joy as they worked.

When it was time to say goodbye, everyone was in tears. Momma had not met Daddy's people before, and she loved meeting his brother and sisters who had come back home to see us while we were there. Jimmy, Momma, and I loved seeing places Daddy knew as a boy, most particularly his fishing hole along the Wabash River.

Daddy loaded our luggage into the car saying, "Come on, kids! We've got a long ride ahead of us!" After we all gave our final hugs and kisses and got into the car, Daddy stepped out one last time, put his arms around his parents, and kissed them.

"Ah, Johnny, I can't thank you enough for all the hard work you did," his mother said in her heavy Scottish burr with tears still streaming down her cheeks. "I just wish ya didn't live so far away, Son. When are we ever going to see ya and your beautiful family again?"

"We'll try to make it back next summer, Ma," Daddy said as he got into the car and started the engine. "Just glad I could finally give you what I promised."

Grandma just couldn't bear to see us leave so she held tightly onto the door handle. "Ah, come on, Ma. Don't make it any harder than it already is," Daddy said with dismay.

"Johnny, take care of yourself, son," she said as she leaned into the car and caressed his face in her frail hands. "I worry about you all the time having to go down into that mine. For God's sake, be careful, Son," she warned. "And, thanks for always sending money home. We couldn't make it without your help, ya know."

"I know, Ma," he replied.

Grandpa finally had to drag her away from the car that day. We kids leaned up against the back seat of the car and waved to them from the back window as we drove away, not knowing if we would ever see them again. And we never did.

CHAPTER SEVEN
ROCK SPRINGS: THE HUB

WHEN THE WAR ended in August 1945, seventy mines operated in Wyoming, employing 4,815 miners. Rock Springs was the hub of the mining communities in southwest Wyoming. It was where all the miners went when they said they were "going to town." For the most part, Rock Springs had everything the mining families needed.

It had the stores the small mining communities lacked: department stores, drug stores, music stores, meat markets, automobile dealerships, gas stations, churches of all denominations, courthouse, hospital, miners' clinic, mortuary, cemetery, hotels, restaurants, lumberyards, movie theaters, train depot, saloons, and brothels.

Rock Springs is not a naturally attractive town, though some would argue. Its barren high desert landscape is a startling contrast to the majestic vistas and breathtaking mountain scenery of the northern part of the state. In fact, some people think Rock Springs is the ugliest town they have ever driven through, though it has improved in recent years. A long-standing joke is: "One of the prettiest sights in Wyoming is seeing Rock Springs in your rearview mirror." But, to me it was the most wonderful place to grow up.

○ × ○ × ○

The families who lived in Rock Springs and in the surrounding mining camps were mostly foreign-born—people who cherished their old country traditions and values. They came from Sweden, Greece, Austria, Hungary, Japan, Poland, Russia, Slovenia, Scotland, England, Czechoslovakia, Italy, Finland and other countries with mining traditions. Fifty-six nationalities, the sign downtown denotes even today.

These immigrants, arriving in the United States by ship, sometimes as stowaways with little money, made their way to the Wyoming prairie looking for work that was familiar. Some women cried for days upon their arrival, praying they wouldn't have to stay long. Their new home left them in shock. It lacked surface water, trees, fertile soil, and had a harsh climate with a short growing season. Everything seemed rough, rugged, and severe. The wind blew every day and the climate was extreme—cold and snowy in the winter and hot and dry in the summer.

But it did have what they came for: the availability of steady, good paying jobs and an opportunity for a better life. Coal seemed to be in unfathomable abundance, not just in reservoirs deep underground, but also on the surface. It burst out in outcrops along cliff faces and eroded to daylight on ridge sides and drainages. Underneath these desolate, wind-swept plains slept "black gold," the likes of which were unimaginable to the new immigrant.

They poured into the Rock Springs area, each nationality called by a unique, derogatory nickname by others, but they all worked and lived closely together. Without even speaking the same language, they formed a community. Coal is what brought all these people together to change the landscape and create villages. Even if you didn't know a single business related to the coal industry, even if you never met anyone whose job depended on coal, in the Forties and Fifties all life in the United States was touched by coal and still is today through electricity.

Rock Springs, with all its ethnic cultures, truly had a unified spirit. Salt Lake City, Utah, 150 miles away, was the closest metropolitan city.

◯ ✕ ◎ ✕ ◯

The people in the mining towns were left to devise their own entertainment. Brewing and bottling beer and wine was one skill the miners from southern Europe brought with them. Far from grape orchards, Italian miners pooled their resources and had a whole car of grapes shipped by rail to Rock Springs and then made homemade brew in their basements. They preferred Zinfandel grapes, which made a deep, red wine they called "Dago Red."

Making beer is a completely different process than making wine. The Polish miners and their wives had the reputation of making the best beer. They kept the beer crocks covered behind the old kitchen coal stoves while each batch fermented. Homemade alcoholic beverages saved a little money and kept the miners away from bars. Families gathered together to visit, play cards or musical instruments like the accordion and piano, pitch horseshoes, play baseball, and tip up a homemade brew or two. Few people drank to excess as the beer was part of a larger social event, though every town has a few notorious drinkers.

In 1913, the Slovenian people constructed a large, unique public hall in Rock Springs called the *Slovinski Dom*, which was used as a gathering place for several Slovenian-American fraternal lodges, a place where Slovenians, Croatians, and Italians hosted meetings, dances, weddings. Later, in the Sixties, it housed various church gatherings as well. Flags of many foreign countries lined the halls. The primary function of the *Dom* was to provide a place where immigrants could feel comfortable with the culture of the old lands while adapting to their newfound home.

Each year in October, the month of the "harvesting of the grape," the Slovenians held a big dance and festival—known as the Grape Festival or Wine Arbor Festival—in the *Slovinski Dom*. The fun-loving people of Rock Springs looked forward all year to this event.

I'd gone to dances at the *Dom* with my parents and grandparents since I was a little girl. My brothers and I loved to run and play in the old building with all the other children, watch the adults dance, eat kielbasa, and drink lots of soda. We'd laugh when we'd see our grandma, who rarely went out of her house, dancing with Grandpa and having a good time.

When I was about twelve, Daddy and Momma let me take my two best girlfriends to the festival. Now that we were practically grown up, we planned to dress up and dance to the live music just like the adults. I wore a nice dress with three crinolines underneath and my newest shoes. I wanted to look nice so I would be asked to dance each and every dance.

We were so excited when we walked into the building. The large dance hall occupied most of the main floor with a draped stage at the far end where the band sat to perform. We scrambled to find a seat on one of the many benches that lined the dance floor. We looked up at the ceiling in awe—a massive amount of fresh fruit, including apples, oranges, and grapes, hung from a large mesh-like screen, similar to chicken wire, that was suspended from the ceiling. It covered the entire ceiling above the dance floor, just high enough that even the tallest men could not easily touch it. Fall leaves, raked from the yards, were scattered in the middle of the dance floor.

After a few awkward moments of feeling bashful and intimidated, my friends and I were soon dancing with boys our age in attendance—some we knew, some we just met. It was the greatest feeling in the world dancing on that dance floor under the grapes and fruit.

In between dances, Momma came up to me and said, "Marilyn, I told you not to wear your new shoes. They're going to get ruined before the night is over. Maybe you girls should take them off and dance in your bare feet."

"Oh, Momma, we'll be careful. We can just wash our shoes off when we get home."

As the night went on and the room got hotter and hotter, the fruit drooped and began to fall onto the dance floor into the leaves. Dancers couldn't avoid stepping on it, or maybe they didn't even try. Soon the floor became a mushy mess, but no one seemed to mind or quit dancing; they just kept dancing in smashed fruit and leaves. When I looked down at my white anklets and black patent leather shoes, they were covered with smashed purple grapes. My girlfriends and I ran to the bathroom, took off our anklets and shoes, rinsed the socks in water, and tried to clean up our shoes. Then we gathered them up and ran looking for Momma. She smiled when she saw us coming and said, "Girls, better put your shoes and socks underneath the chairs Grandma and Grandpa are sitting on. It is okay to dance in your bare feet tonight; everyone else is."

So we danced barefooted as the band, made up of local musicians, played popular polka, waltz, and jitterbug music. Except for the piano, which belonged to the hall, the band members brought their own instruments: accordions, drums, saxophones, and clarinets. Some couples danced so well that everyone else stopped dancing and stood admiring the talented couples twisting and turning to the beat of the music.

I always loved to see the many foreigners dressed in their traditional old country garb, especially the older people who took great pride in their country of origin. But, I have to admit, I was a little shocked when I saw women dancing together. Momma said they were women who didn't have a husband or had a husband who did not dance. They wanted to dance badly enough that they danced with each other.

The dance floor was crowded all night long: young adults dancing with their sweethearts or friends, children dancing with other children or parents, and grandparents dancing with their spouses or friends. As a very young girl, I loved it when Daddy came to get me to dance with him even though I had to stand on top of his shoes to do it. But, when I got older, he danced with me just like he did Momma, and I loved it. He was one of the best of the dancers, and he taught me special steps.

The Grape Festival seemed to release people of their inhibitions, and they pushed onto the dance floor whether they knew how to dance or not, smiling, laughing, and having a good time. Many times the whole crowd sang to the songs being played. It was truly an unbelievable feeling—belonging to a joyous, sweaty tribe of all generations, caught up in the moment and the music, celebrating without worrying about tomorrow.

○ × ○ × ○

The whole time we were there, wonderful smells drifted up from the kitchen downstairs. Food and beverages were sold in the basement for a minimal price. The star of festival food was smoked garlic sausages, called kielbasa, which were simmering in large kettles on the big coal stoves. Everyone ate them on hotdog buns, garnished with ketchup or mustard and sometimes sauerkraut and washed down with soda pop, beer, wine, or hard liquor. When we first arrived at the *Dom*, Daddy reached in his pocket and gave us kids money so that whenever we got hungry, all we had to do was run downstairs and buy whatever we wanted. We made several trips downstairs until the money ran out; then we looked for Grandpa who gave us more.

Everyone's favorite Slovenian dessert was a sweet bread called *potica*, the same bread that earned Ann Cummings a reputation at the boarding house. Women of the community prepared this in their homes to bring to the festival, resulting in an undeclared cooking competition. I always tried to have a slice of the *potica* my aunt Edna Copyak made, and the servers knew the very loaves she had brought. The sweet dough was rolled very thin, covered with a buttery, crushed-nut, sugar, and egg filling, and then rolled into a jellyroll-type loaf. Baked until the crust was a rich, dark brown (as only a coal stove could do) and set on racks to cool, it was served plain or with a buttery frosting on top. Almost everyone in attendance had at least one slice of the delicacy. Of all the special things from my childhood, the Grape Festival was one of my favorites. I attended those dances every year until the day the building was closed.

○ ✕ ○ ✕ ○

Rock Springs was special to me for another reason; my grandparents lived there. While Rock Springs was the hub of the coal communities, my grandparents' house was the hub of our family life.

My Rock Springs grandparents, my mother's parents, came from Czechoslovakia. During this time every family identified itself by a distinct nationality, and most immigrants tended to marry within their own nationality. Daddy's parents were from Scotland and, after he and Momma married and my brothers and I were born, our family became a mixture of nationalities for the first time. We ate many old-country Czechoslovakian dishes prepared the traditional way, observed many of the customs, and even tried to learn to speak the old language with the encouragement of our grandparents. Many times my grandparents referred to people by their country of origin, rather than their last name, and we all knew exactly whom they were talking about. For example, grandfather would say, take this to "the Finnish man down the street" or "the Italian girl who lives with her mother." We didn't consider it a prejudice to refer to a nationality; it was simply a way of identifying people, similar to saying "the tall man who lives on the corner."

Many families spoke their native tongue in their homes, but quickly learned English so as not to appear foreign when out in public. Some even had their last names changed to sound more American. Religion was a part of the cultural differences of some nationalities.

My grandparents were Catholic and every night, before going to bed, my grandfather would kneel before a picture of the Virgin Mary

Grandma and Grandpa Copyak (John and Susan) posed with their new Ford in the 1940s.

and pray. He called me Mala, rather than my given name, and on nights when I would stay over at their home, I shared that special moment of prayer with him. Right before bedtime, after I put on my nightgown, Grandfather would call to me from the dimly lit dining room where the religious picture hung, "Come, Mala. Fall to your knees. We pray together to our Lord."

Even though Grandfather had gone to school in Czechoslovakia only through the third grade, he learned to speak fluently and write in four different languages: Czech, English, Russian, and German. He taught himself from books borrowed from friends and neighbors and by learning from people he knew. Many a night he huddled over the dining room table filled with books, his glasses down on his nose, studying intently, like a college student cramming for exams. He taught himself not only languages, but also basic math through algebra. He constantly acquired books and studied to become more knowledgeable, probably because he hadn't had the opportunity to attend much school as a young man. When he was thirteen years old, he began working in the mines to help support the family who took him in when he came to America.

○ × ○ × ○

Many a night when I stayed at my grandparents' house, I greeted neighbors who tapped on the door in the evening to ask grandfather to read letters they received from the old country. I sat with my elbows on the dining room table supporting my head, listening and watching as he carefully read the letters aloud, sometimes in a language I could not understand and sometimes in English. I watched tears overflowing the eyes of the visitors as they heard news and wondered about the children or parents they'd left behind. Then he helped them with a reply, sometimes with his perfect penmanship and sometimes by typing on an old manual typewriter. The visitors handed him fresh garden fruits and vegetables or shyly left homemade foodstuffs on the kitchen table. He accepted gifts graciously, but never accepted money. Even as a child I recognized how much he valued knowledge, and I was amazed at how well he had taught himself. I was proud to be the granddaughter of this man so respected for his learning.

He often talked about life in the old country. His father had died, and his mother's boyfriend did not want to marry anyone with small children. In 1905, at his suggestion, his mother decided her son should go to America—alone. He was only thirteen when his beloved mother and her boyfriend took him in the dark of night to board a freighter bound for the United States. He described how he stood waving goodbye to his mother for the last time, until he could no longer see her and long after she must have left the dock. He also told of the hardships he endured on the ship and what it was like, alone in a new country when he was so young. Fifteen cents was all the money his mother had given him. With that, he bought a cherry pie when he got to New York City.

He told me how appreciative he was of the family who found him on the streets of New York and took him in, and how that family led him to find work in the coal mines. He was grateful to the Union Pacific Coal Company, which he held in the highest regard, for giving him his first job, even though the work was backbreaking and dangerous. Often he would cry as he recalled the stories from his past. I couldn't imagine that a thirteen-year-old boy would be sent to America, not speaking English, with so little money, and be expected to survive.

○ ✕ ○ ✕ ○

Grandpa was the stalwart patriarch of the family. When Nikita Khrushchev declared "we will bury you," the whole country became afraid of Russia, especially us kids. We all worried, but Grandpa took action.

Grandpa began digging a bomb shelter in his backyard, as big as a foundation of a house. And he did it alone while many of his neighbors laughed at him. On weekends when family members sat in his yard drinking beer and conversing, Grandpa shoveled dirt. Daddy picked up a shovel and helped him during each of our visits. They framed in the underground sides of the shelter, and stocked it with food and water in five gallon containers.

Years later, my brother Johnny stayed with Grandpa and loved using the bomb shelter as his private apartment, though there was no electricity and only cold running water. Johnny used a propane camp lantern for light, turned on his battery-powered transistor radio for music, and even took cold showers

In 1968, Grandpa became understandably fascinated with the episode of Lloyd Bucher, who was the adopted son of Grandma's sister Mary. Bucher was the commander of the surveillance ship the *USS Pueblo* that was hijacked on the high seas by Koreans during the Cold War era. For almost a year, radio and TV news reported daily on the status of the eighty-three-member crew.

Grandpa told me the story of how Grandma's sister Mary came to adopt Lloyd. Aunt Mary and her husband, Austin Bucher, moved to Pocatello, Idaho, during the Depression, where they owned a restaurant. Aunt Mary regularly visited a friend in the hospital, and each time, she heard a baby crying in the nearby hospital nursery. The nurses said the baby was the child of a Nez Perce woman and a Swedish man and was available for adoption. Mary fell in love with the baby from the first time she saw him, and she and her husband adopted him.

When Lloyd was three years old, Mary died of cancer, and before long Austin lost the restaurant. He struggled to find work and raise Lloyd with help from his mother. But when his mother died, Austin didn't know what to do with the boy, so a sister of Aunt Mary and Grandma's, Ann Rap, agreed to help raise him. Ann and her husband lived in Glendale, California, so the adults shuttled Lloyd back and forth between Pocatello and Glendale. Lloyd became unruly as he grew, and Austin Bucher and Ann Rapp found it hard to manage him, so they placed him in an orphanage, which segued into a series of orphanages.

Through it all, Lloyd loved to read and write. When he was fourteen years of age, he wrote a letter, with the help of a nun, to Father Flanagan of Boys' Town. Flanagan accepted him to the facility in the summer of 1941, where he excelled in academics and sports until he dropped out his senior year to enlist in the Navy.

After he got out of the Navy, Boys' Town officials helped Lloyd get a football scholarship to the University of Nebraska. In 1954 he was called back to active duty, and in 1960 he was assigned to the *Pueblo*.

One summer my family vacationed with my Aunt Kay in Redondo Beach, California, and Lloyd was visiting there at the same time, on leave from the Navy. I was around thirteen years old and thought Lloyd looked just like Robert Wagner, the actor. We spent many hours swimming together in the Pacific Ocean. Grandpa was proud of Lloyd's accomplishments in spite of his early hardships.

Lloyd later became the commander of the *USS Pueblo* and stood at the center of one of the most incredible events in the history of the U.S. Navy. He and his crew endured brutality at the hands of the Koreans in an attempt to get the captives to confess to violating territorial waters to spy on North Korea.

My family gathered at my grandparents' house to watch television on the night the news first showed images of the captured *Pueblo*. Grandma cried as she watched for her sister's son and prayed that he would be allowed to come home. It was a very tense time for our family.

After eleven months of torture they were released and, one at a time, they walked across the border into South Korea while our family watched for the familiar face of Lloyd, cheering when he was safe.

CHAPTER EIGHT

MAKING THE BEST OF SUMMER

AT THE BEGINNING of each summer, Momma did something that entertained my brothers, me, and the other children in the neighborhood for weeks. She went to Rock Springs and bought enough small cars and trucks from various stores to fill a huge cardboard box. With those cars as props we spent hours turning our backyard into a giant city of mud houses and highways. We discovered a gray-colored dirt in the surrounding hills which when mixed with water made a clay-like substance perfect for paving streets and highways in our pretend city. We hauled bucketloads of this dirt from the hills, added water from our backyard hose, and built miniature yet realistic highways, streams, and dams. Momma said it was worth it to buy the cars just to keep us kids at home, so she didn't have to worry about where we were.

The summer after I graduated from elementary school, we once again began working on this backyard project when Daddy announced during dinner a few projects of his own. Momma smiled all over when he said he was going to build the white picket fence she had wanted for so long and then plant a lawn and garden. "Who wants to give me a hand?" he sang out with raised eyebrows, and we kids shrieked, "I do, I do!"

"Well then, I'll go to Kellogg Lumber in Rock Springs on Saturday and order the lumber. I'll also find some good topsoil for the yard, once we finish the fence."

On Saturday, our parents dropped my brothers and me off at the Rialto Theater in Rock Springs to attend the matinee while they drove around town stopping to pay bills, buy groceries, and shop at Kellogg Lumber. Then Daddy took us to dinner at the new diner located in White

Mountain addition. An old passenger train dining car had been set on a concrete foundation and converted into an eating establishment owned and operated by a local family. A huge, fancy neon sign graced the top and bathed the area in flashing colored light. On the inside of the diner, a long counter ran along one side and tables occupied the other side. A jukebox sparkled from the end wall, and small control boxes hung from the wall at each oilcloth-covered table. I begged for a dime to insert in the coin slot, and Daddy let me select my favorite songs. It seemed like he knew and visited with every customer who came in.

○ × ○ × ○

Our family spent evenings and weekends that summer working on our new yard. Occasionally friends and neighbors pitched in, and in no time we'd finished building the fence. Daddy allowed my brothers and me to help paint the fence with two coats of white paint. Momma could hardly wait until we could get trees and grass planted, not only for the beauty, but also because a lawn would cut down on the dirt tracked into the house.

When the fence was completed, Daddy surprised everyone when he backed the big company dump truck into our driveway carrying a swing set that one of the men at the mine shop had made. Every kid in the neighborhood went wild. Except for the swings on the school playground, we now had the only swing set in camp. We jumped up and down in anticipation waiting for Daddy to unload it and put it together as the crowd of our young friends grew. From that moment on, all the kids in camp played in our yard, and Momma and Daddy liked it that way. They never told kids to go home and play in their own yard like some parents did. Momma wasted no time in sending some baked goods to the man in the mine shop in appreciation.

Daddy still had the dump truck the following weekend, so Momma packed a lunch, and we went on the first of many trips to a place our parents sometime took us for picnics south of Winton. Daddy let us ride in the back of the dump truck with our dog, Tippy, if we promised to sit down and not hang over the sides. We headed toward Winton, singing all the way, as we bobbed along in the back of the truck. Tippy loved the ride and stood stately with the wind blowing through his fur.

My brother Jimmy Nesbit, Sharon Logan, Georganne Pecolar, Carolyn Pecolar (seated) and I play on the swing set in front of Stansbury school.

Daddy stuck his head out the driver's side window and yelled back, "You kids sound like a bunch of Chinamen with all that singing!"

Daddy stopped the truck outside of Winton near a creek lined with hundreds of tall pussy willows. "Come on, kids. This spot probably has the richest soil in all of Wyoming," he proclaimed. My brothers and I jumped from the truck and ran through the willows, feeling we could easily get lost among them. Daddy arduously loaded the topsoil, throwing one shovelful after the other up into the bed of the truck as sweat ran down his face. It took him the better part of the afternoon to get a full load; he stopped only to eat the lunch Momma had packed for us. When it was time to leave, he summoned us by whistling.

Daddy threw his shovel on top of the soil and told us to grab the dog and squeeze into the cab. Just riding in that big truck, sitting up so high with all the windows rolled down and the dog hanging his head out the passenger side window, was so much fun. Back at the house, we sat in the truck to watch Daddy throw the switch that dumped the dirt right into our yard. Then we scampered out to help spread the soil with shovels. Daddy filled the wheelbarrow and scattered topsoil into the area that would become his garden.

After making a few more trips for soil over the next few weekends, Daddy went to the house we had lived in at Winton, which was now vacant, and dug up the trees to transplant into our yard in Stansbury. By the end of the summer, we finally had a yard filled with trees, grass, flowers, and what was to become a thriving vegetable garden. Daddy made a wooden lawn couch and chair which he painted white with the leftover paint from the fence. Now, more than ever, kids clamored to play in our yard.

○ × ○ × ○

When Daddy was a young man back home in Indiana, he prided himself in his ability to grow things, especially tomatoes. People in Stansbury found it almost impossible to grow vegetables, but Daddy decided to give it a try. He loved fresh tomatoes. When he could harvest enough tomatoes, he made a special piccalilli relish from his mother's recipe.

He meticulously planted and babied his garden, checking on it the minute he got home from work each day. Momma said it was his "quiet time," when he got away from all the noise and problems he'd had at the mine. The sight of him enjoying himself, watering his garden with the wind slightly blowing through his hair, is a picture I hold in my memory.

○ × ○ × ○

Most of the kids in camp took to the hills surrounding Stansbury on summer days. We hauled scrap wood from our yards and gathered flat rocks to build "cabins" on the hill near the camp where the water tower stood. Once the cabins were constructed, we built a big spit for bonfires so that we could roast wieners, potatoes, and marshmallows. On each trip up the hill, we brought more and more things from home: newspaper, matches, materials to create makeshift benches, old blankets to put on the dirt floors of our cabins. Unbeknownst to us at the time, our projects were under the eye of the camp watchmen who patrolled the camp and surrounding hills day and night.

Thank goodness a watchman was there the day we really needed help. One afternoon while we were playing around our cabin, someone shouted, "Let's go swimming in the water tower!"

I was quick to remind my brothers loud enough for all to hear that Daddy had warned us never to go near the water tower.

"Ah, come on. Let's climb up the ladder and just look in," someone wheedled. It didn't take long for the adventuresome bunch to head up the tower ladder. I stayed on the ground because I was afraid of heights. It made my heart race just to see the boys, including my brother Johnny, climb higher and higher up the ladder.

The next thing I knew, they disappeared over the top of the tower. I visualized them climbing down the short ladder we all knew was inside the tower, then, when they reached water level, pushing off to swim around. On the ground, I listened with envy to their raucous shouting and laughing.

Within minutes I heard someone yell, "Let's get out of here! The water's going down." If the water went down, they would no longer be able to reach the short inside ladder to climb out. If trapped in there long enough, the boys would tire and drown. I ran down the hill toward camp and saw a UP pickup truck headed my direction. I waved my arms above my head and screamed, "Help, help. The boys are in the water tower. They're gonna drown." The driver, who happened to be a watchman, rushed to the tower, climbed up the outside ladder with a rope hanging from one shoulder, and, one by one, pulled the boys out.

When everyone was safely on the ground, the watchman couldn't contain his fear and anger. "You damn kids shouldn't have been up here. What would have happened if I hadn't come along? Ya always gotta think before you do things! I'll have to tell all your parents about this, hear me? Now get the hell home!"

Wet, cold, and scared to death, we headed for home. That night at the supper table to our relief, everything seemed normal when we sat down. Suddenly Daddy slammed his fist on the table. We jumped in fright and, as much as we wanted to run, we sat tight. He said, "Margaret, the kids could have drowned today. I warned them never to go up there, but they didn't listen."

We knew all what was coming. He ordered us to go into the living room where he lined us all up and whipped our bottoms with his belt. Then he sent us to bed without supper while it was still light outside and grounded us for two weeks. Licking our wounds, we cried ourselves to sleep behind the closed bedroom door, sure we were starving to death.

○ × ○ × ○

Most of the kids in camp joined all of our adventures. However, a couple of families would not allow their children to leave their yard or even their house. One little girl, not yet old enough for school, whose parents were older than most, never left their house. Not only did she have to stay indoors, but her mother kept the blinds in the house closed. One day we glimpsed her peering out from the side of the closed blind. Shocked to see how pale she looked with very dark circles under her eyes, I told Momma what we had seen. She told us not to criticize how other people lived and to stay away from their house.

Some mothers would not let their children go out to play until they had finished all their chores. I once told Momma that some of my friends had to do the ironing, wash a week's worth of dishes, clean the house, and stay home to babysit their younger brothers and sisters while their parents went fishing or into town to the bars. My exact words were, "Their parents make the kids do all the work. They're mean and lazy and drink too much." Momma lowered her head and told me not to be concerned about what other people did—it was none of my business.

I can remember going over to one friend's house to see if she could go to the movies with us. Her parents weren't at home, but she said if she got all her work done maybe her mother would let her go, so I volunteered to help. While we mopped the bedroom floors, she said, "Want to see something?"

Of course, I did.

"Come look in my mother's dresser drawer, but we have to be careful so she doesn't catch us."

Cautiously looking all around, she pulled open the drawer and revealed all kinds of candy bars. I asked if we could take one. "Oh, no! She would know if even one piece was missing, and I would get a beating."

As we closed the drawer, her mother's harsh voice came from behind us. "I thought I told you, young lady, never to look in my drawers. Marilyn, go home and don't ever come over here again."

As I walked out the back door, I could hear her mother beating her with a belt, and my friend sobbing and pleading, "Oh, Momma, stop! Oh, Momma, stop! I promise I'll never disobey again."

My brother Jimmy and I proudly stand by our new Schwinn bicycles purchased at the UP store in the 1950s.

I felt miserable as I walked home thinking about her. She had to work so hard. And wherever she went, she always had a younger brother or sister hanging onto her or in her arms. Things never changed at that house over the years. Often when I passed by her house, day or night, I was saddened at the sound of my friend crying. It was so bad at times that I'd cover my ears and run the rest of the way home. I saw the black and blue bruises on my friend's arms and legs when she came to school, the only time we played together.

One afternoon she and I played jacks on the school porch and she confided, "One of these days you won't see me anymore. Just as soon as I'm eighteen, I'm going to run away, and I'm never coming back. Never! I hate my mother!"

I didn't know what to say but I could tell by the look on her face, she meant it. Then one day her family moved, vanished almost overnight. No one seemed to know why they left, only that they were living somewhere in Rock Springs. Their house sat vacant for a little over a month, and I couldn't help but think about her each time I walked by.

○ × ○ × ○

One evening each summer Daddy would announce to everyone playing in the yard that he would repair any bikes that needed fixing. In no time, the yard filled with bikes. He tightened bicycle chains, handle bars, baskets, and patched tires. If there was nothing the matter with a kid's bike, the owner sometimes looked for some little thing that wasn't right, so he could have his bike in the line for repairs. Once all the repairs were completed, Daddy oiled each and every bike.

○ ✗ ○ ✗ ○

Our telephone system was a party line with as many as ten families on one line. Our phone was a rotary model. Momma told us if we wanted to make a call and we discovered that someone was already talking on the phone, we should hang up immediately. She said if we had an emergency, to ask for the line politely, and people would wrap up their conversations.

Most people were very nice about restricting their time on the phone, but, of course, everyone on the party line could listen in to the conversation if they wanted. Daddy limited the time my brothers and I spent on the phone. He told us the line had to be clear in case the mine called.

○ ✗ ○ ✗ ○

One of the camp kids' favorite places to explore was Red Rocks, about a mile south of camp. When we decided to play at this rock formation, we took lunches. We hiked up the dirt road leading to these giant red rocks with sand at their base. We took off our shoes to feel the hot sand on our feet and sometimes pretended we were at a beach on the ocean. When we tired of that, we put our shoes back on to explore. Up in the rocks, a maze of caves led out onto high ledges where we could look out across the plains and see Stansbury far off in the distance. We felt like we were on the top of the world, with the sun beating down on our bodies and the wind blowing against our faces. Kids covered the rocks like ants, each one exploring newfound places.

On some occasions, with our parents' permission, we stayed at Red Rocks until sundown, built a bonfire, and roasted potatoes and marsh-mallows. We were never afraid to be there without our parents or to walk home in the dark. We knew that if we weren't home by ten, the camp kids' curfew, someone would come looking for us. In all the years we explored Red Rocks, I never remember anyone getting seriously injured.

○ ✗ ○ ✗ ○

The camp kids looked forward to participating in the summer recreation program funded by UP. The company hired a couple of teachers from each camp to direct a summer program held at the school—arts and crafts classes, well-organized baseball teams, and a swimming program. The buses from Reliance High School transported the children to and from the different camps for baseball games. On Fridays, the buses took everyone to Rock Springs where we swam in the new Rock Springs High School pool. This entire program was free to all the miners' children.

For years, Mr. and Mrs. Lanoy were in charge of the summer program at Stansbury. The Lanoys were an extraordinary couple who didn't have any children of their own and seemed to devote their every waking moment to working with other people's children. Many times on a Saturday, the Lanoys drove a carload of children to Steele's Hot Springs in Thermopolis, to swim in the indoor, naturally-heated mineral pool. The town of Thermopolis, a three to four hour drive from Stansbury, was known for its many of hot mineral springs. Enterprising local folks had created covered pools, popular with kids and adults alike. We left early in the morning and got home long after dark.

I kept in touch with the Lanoys, whom I loved like second parents, long after I left Stansbury. Once, long after Mr. Lanoy had retired, I asked him which school was his favorite. Without any hesitation he replied, "Stansbury was my first and my favorite—mainly because of the people. There was just something special about each and every one of those families, and the kids were so eager to learn, and so appreciative of everything that was done for them. They were fortunate to have wonderful, caring parents."

o x o x o

Summer was also the time when the UP painters painted all the camp houses, probably a necessity to keep the camp looking good in spite of all the coal dust. One year they painted the inside of the houses and, the next, the outside. No one had a choice of colors; everything was painted white. If someone wanted rooms painted a different color, they had to buy their own paint and do it themselves.

In the summer of 1953, the painters arrived to paint our house on the outside. My brothers and I gobbled down breakfast and hurried outside to await the painters. The senior painter drove up in the oldest car I had ever seen, but it was so well cared for it almost looked new. When he got out, he wiped the dust from the bumper of the car, which shined in the morning sun. They unloaded paint cans, tarps, and brushes while we asked them a million questions. They responded with frowns and grunts and groans, probably wishing Momma would call us back into the house so they could get on with their work. After a while, Johnny and I got bored and played other things in the yard, checking back occasionally, but Jimmy stayed, volunteering to do anything to help. He was dying to get his hands on a paintbrush.

The boss painter wasn't much for conversation. He told Jimmy time and again to get away from things and go play with the other kids. Quietly Jimmy sat down on the porch steps and watched the men paint the back side of the house and then move their equipment around to the side, whereupon he followed them. He asked once more if he could help.

"I told you, kid, get out of here, so we can get your house painted. You're driving me nuts," the older fellow snarled. "Now, I mean it, get out of here!" With that, he walked over to Jimmy and smacked him on the top of the head with the paintbrush he held in his hand. White paint filled Jimmy's hair and ran over his head. Paint ran down his forehead and gathered a moment on his eyebrows before dripping over his eyes. Streams ran into his ears and down the bridge of his nose, where it dripped off to his chin. He ran screaming into the house, making things worse by wiping at the paint with his hands. Johnny and I followed to see what would happen next.

Momma was washing clothes in the basement and immediately ran upstairs as Jimmy screamed bloody murder, "Momma, Momma . . . look what that mean old man did to me!" With his voice quivering he shrieked, "I can't see! I can't see! My eyes are painted shut."

Momma grabbed Jimmy by the arm and promptly ran out the back door to confront the painter. After a rash of heated words on both sides, she went back into the house, ran downstairs to turn off the washing machine, then back upstairs to begin cleaning the paint off my brother with turpentine, muttering, "What kind of crazy man would do something like this to a little boy?" Momma was horribly upset as she worked to clean off all the paint, irritated at the insult, the mess, and how the incident had disrupted her busiest workday. Johnny and I were ordered back outside.

After what seemed like hours, Jimmy emerged from the house with most of the paint out of his hair and ears. Again he sat quietly on the front porch with his arms resting on his knees holding up his face. His face was beet red from all the crying and scrubbing. Then he got up and went to the side of the house.

About five o'clock the mine whistle blew, and the painters gathered all their equipment to go home for the day. When they rounded the corner of the house to where their vehicles were parked, an amazing sight lay before them. The grumpy painter's old car had been given a new coat of white paint: bumpers, windows, tires, and all. Johnny and I and several neighborhood kids had gathered in anticipation of what would happen next.

The painter stood as if turned to stone with a look of hatred on his face I'll never forget. He sputtered and took a step toward Jimmy. We all stepped back.

Momma came out the back door to take trash to the outside garbage can and noticed the silent crowd at the side of the house. She'd regained her good spirits and sang out in a melodious way, "What's going on, kids?" as she walked over to where everyone was standing. Her eyebrows rose until they almost disappeared into her hair. Her mouth fell open. She dropped the trashcan from her hands. "Oh, my God," she whispered.

The old man angrily yelled, "Just look what that damn kid of yours went and done to my car! And, what the hell are you going to do about it?"

Momma took a breath and her eyebrows came back down. Then she took another breath and threw back her head. "Not one thing," Mother said with her chin held high. "You whacked your paintbrush over my son's head, and I had to stop everything to clean him up. You disrupted my whole day. So, the way I see it we're even!" Whereupon she took Jimmy by the hand and together they marched back into the house.

I was never so proud of Momma as I was on that day. The soft-spoken woman I had always known her to be had been pushed to her limit, and she stood her ground. The next day the painters returned to complete their job. We watched from around the corner of the house as the grumpy old painter drove up in his newly white car. He stooped forward in the car seat to peer through the spot in the windshield he had cleared of paint. But his attitude was different, more subdued, even to us kids.

Daddy had unexpectedly been called to a meeting the night before. We'd worried and had a hard time sleeping. Was he having to answer for our actions? Would Jimmy be sent to reform school? But that morning nothing had been said. Nothing at all.

After a few days of skirting around our parents, we surmised that the painter had been on the receiving end of one of Daddy's "little talks." Or maybe it was Momma who had caused the painter's change of attitude.

CHAPTER NINE
HAUNTED BY THE WHISTLE

I KNOW NOW that life dissipates as easily as steam from a coal locomotive, and each tomorrow may be anticipated but is never guaranteed. We coal camp children understood this on a simple level—because of the mine whistle. If we heard the wail of the whistle at an unscheduled time, no matter what we were doing, the bottom dropped out of our stomachs and sometimes the bottom dropped out of our worlds. All of our fathers and almost all the adult men we knew went down into the mine at the beginning of a shift and, though we anticipated that they would emerge again, sometimes the plaintive sound of the whistle flowed down our streets, around the corners, into our homes, and pierced our hearts.

We children could list the common types of mine accidents. Sometimes at night before we slept we thought of explosions and cave-ins; methane gas pockets that poisoned miners or sparked into fires; oxygen shortages; and all the dangers of working with explosives to blast coal away from the mine face. While other children lay sleepless thinking of bogeymen under the bed, we coal camp kids were haunted by the sound of the whistle and what it might bring.

The whistle was the regulator of life in the coal camp. It routinely signaled the beginning and end of each shift, and it sounded for a national emergency or celebration. But the sound we dreaded was when it blew incessantly at a time we were not expecting it. That signaled a mine accident. Young and old stopped what they were doing and ran. We ran toward the sound, toward the mine portal, praying that the accident was minor and that our miner was safe.

The monthly *UP Employees' Magazine,* published by the Union Pacific Mining Company, was filled with camp happenings and company news along with recipes, births, and deaths; but the section that made us want to look away showed the number of mining accidents in each camp mine, listed by the month and the year.

Mine injuries ranged from burns to severed limbs and paralyzing spine injuries to death. Sometimes after the long whistle had blown, other miners gathered at our house after work. Daddy asked us kids to go outside or to our rooms to play. We knew then that something really terrible had happened. We tried to listen and overheard stories of bravery, of injured miners who walked out of the mine, bleeding and in pain, rather than waiting for the mantrip to transport them to the waiting ambulance.

One story that has always stuck in my mind is the day Johnny Bozner's hand slipped as he was working on a duckbill (although I never knew what this was) and his hand went into the machinery, smashing all his fingers. His wife, Evelyn, repeated the story of what happened as she heard it from her husband. Several miners quickly bandaged Johnny's smashed hand, using a "cravat" cloth from one of the first aid boxes positioned throughout the mine. Johnny walked out of the seam until he got to the mantrip which took him to the surface. He refused any assistance getting into the ambulance. On the way to the hospital in Rock Springs, the ambulance broke down and, even though Johnny must have been in severe pain, he struggled out of the ambulance and helped the driver fix the motor so they could continue to the hospital.

At the hospital, Dr. Krueger operated on Johnny's hand, which was an unrecognizable mass of crushed bones, skin, and tendons. He tried to heed Johnny's pleas to leave as much of the fingers as he could, so he could return to work at the mine, the only job he knew. The doctor was successful.

Then there was the time my Uncle George hired on at the Stansbury Mine the day after he returned from Germany at the end of World War II. It was his first day of work underground. While shoveling coal in the dark mine shaft, he stepped on the teeth of a rake that someone neglected to stand against the wall. The handle flew up and hit him in the face, shattering his safety glasses. With splinters of glass stuck in one eye and blood all over his coal-dust covered face, he struggled to see as he walked out of the mine to the ambulance. When the doctor met him at the hospital, he took one look at him and uttered, "George, this doesn't look good."

"Doc, for God's sake. You got to save me eye," Uncle George pleaded. In spite of many surgeries, he never regained sight in that eye, but he did recover enough to go back to work at the mine.

I remember only two times when Daddy sustained an injury, and he never missed a day of work as a result of either of them. The first was when a conveyor belt caught his leg, cutting a big piece of flesh out of the calf. He didn't even go to a doctor. He just came home, cleansed the wound, covered it with Cloverine salve, and wrapped a cloth bandage around it. The wound healed itself, leaving only a slight indentation in his leg.

Another time I overhead Daddy telling Momma about catching a unit foreman from a different seam marking his own seam's number onto the coal trip cars that Daddy's men had loaded. The miners prided themselves on how many cars their seam loaded each day, marked them with chalk, and received credit at the end of each shift. Daddy grabbed the unit foreman by the shoulders with both hands and threw him to the ground, warning, "I ought to beat the hell out of you for this, you son of a bitch! So help me God, if I ever catch you doing this kind of shit again, I'll kill you."

The unit foreman didn't move or utter a word as Daddy walked away. The miners who witnessed the confrontation glared down at the offender with their shovels swung over their shoulders ready to take a swing at him if he made a move toward Daddy. That unit foreman never regained the trust of the other miners.

Daddy's words were repeated around the camp over and over as a sign of Daddy's loyalty to his men. He didn't let anyone cheat his men.

Daddy, still furious over the incident when he boarded the mantrip at the end of that day, inadvertently stood up for a moment while the mantrip was in motion. His head slammed into one of the overhead timbers leaving a big, red bruise across his forehead. We were all worried when we saw the black and blue bruise on his forehead. Momma wanted him to see a doctor, but all he asked for was a couple aspirin and a cold glass of water. He said he had a heck of a headache and wanted us kids to be quiet so he could sleep. Then he went directly to bed without eating supper. The next morning he got up and went to work as if nothing had happened.

I only saw Daddy cry twice. Once in joy when we visited his parents and once in despair after he came home from work one day. We knew something was terribly wrong that day. He went into the living room and sat down in his favorite chair, but rather than pick up the newspaper, he just sat there staring straight ahead. Momma sat on the arm of the chair and quietly asked what was wrong as she put her arm around his shoulder. Daddy put his head into his hands and began to cry. "There was a hell of an accident down there today. One of my men had his leg cut clear up into his hip. The mine car he was riding in ran through an open switch and collided with another car."

The injured man was Evan Reese, a "nipper rider" on the motor-driven car. A nipper threw the switches for the underground coal cars to switch to the different tracks. As many as forty or more five-ton coal cars were often down in the mine, transporting empty or full cars, on the trolley-like mine system. Daddy said if that ever happened to him, he'd rather be dead. He never wanted the mine "to put me in a wheel chair."

The room was still until my little brother, with his head lowered, mumbled, "I hate that mine! I want our family to move away from here." After that, no one said a word the rest of the evening, even at the supper table. We kids sprawled in front of the radio listening to our programs, Momma sat on the couch crocheting, and Daddy sat quietly, deep in thought. We had never seen Daddy like that before, and the thought of something so powerful that it could overwhelm even our father left us feeling like all the things we loved in our life were uncertain and drifting.

Yes, mine accidents, and there were many, were an inevitable part of life in a mining community. Whenever there was an accident, the State Mine Inspector invariably seemed to rule in favor of the company and concluded the miner was at fault for not taking proper precautions. The community's response was always the same: banding together to offer help however they could, be it with food, money, or whatever was needed.

o x o x o

One hot afternoon a bunch of us were riding our bikes around camp when someone shouted out, "Does anyone want to go swimming?"

"You mean in Rock Springs?" I asked.

"No. In the creek between North and South Camp," was the reply. "The water coming from the mine is really deep today." This was a place we'd never thought of swimming before so we all raced home to put on our swimming suits under our clothes, grab towels, and hurry to meet others along the creek bank.

The creek was always filled with water coming from the mine, and on that day the murky, reddish brown water came rushing down from the mine faster than normal. "We got to dam it up if we want to swim in it," someone yelled out as we stood on the bank among the big sagebrush, looking down at all the water. We slid down the embankment and began building a dirt dam with our hands. A couple of the kids ran home and brought back shovels to speed up the project, while others gathered up big rocks for reinforcement.

Once our dam was complete, we stripped off our outer clothes down to our swimming suits and jumped into the water, which by then was about five feet deep. "Don't put your heads under the water because it tastes like a rusty old pipe," someone alerted.

"Hey, you guys, look what the water did to my body." I climbed out of the water. I was covered with a gritty, reddish-colored mud. "How are we going to wash this stuff off?"

"We'll sneak into the bathhouse before the miners get there and run through the showers before we go home. You'll see, it will come off," someone assured me. We swam in the creek for hours and then gathered up our clothes and towels and headed for the bathhouse, dressed only in our swimming suits, covered completely in mud. We sneaked into the shower room, laid our things on a bench by the back wall, found one bar of soap and then turned on all the shower heads and proceeded to run through the shower heads washing the dirt off our bodies. "This was the greatest day ever, wasn't it, Marilyn?" my girlfriend sang out.

"I can hardly wait until we go there again," I replied with enthusiasm. We wiped ourselves dry and put our clothes over our wet swimming suits and headed home.

We continued to swim in that creek until the day the UP found out about our secret swimming hole and sent men to break the dam. Our parents were furious when they learned what we had done. We had a family meeting that night where Daddy was very serious when he said, "I just cannot impress upon you enough how dangerous it is for you kids to swim in that creek filled with water coming from the mine. It's full of all kinds of chemicals that could have made you deathly sick. The watchman who broke up the dam said the water was almost six feet deep in some places. Someone could have drowned. Now I don't want to hear any *ifs, ands,* or *buts* about it. You are forbidden to ever go near that creek again. Do you hear me?" he asked as he gave a stern look to my brothers and me. "And, if you disobey me on this one, you will be confined to the yard for a month, with no friends allowed over." Needless to say, we never swam in the creek again.

I've sometimes wondered if we damaged our long-term health that summer, if my friends suffered from more cancers or Parkinson's or MS as they aged. I don't know the answer, but for the short term we loved those summer swims.

CHAPTER TEN

AWAY FROM COAL

Jack Hensley lived in the house right next to ours. He was a widower, raising two teenaged children. Most of the time, Jack was grumpy and hollered at the children in the neighborhood whenever they passed his house. But, on Halloween night he would answer his door and hand out a generous number of coins rather than candy without saying a word.

One year rumors spread around camp that Jack was about to get remarried. While our parents talked about how happy they were that Jack had finally met someone, all we kids could think of was how sorry we were for the woman who would have to live with him. We plotted with our friends to get even with him for hollering at us all the time by chivareeing him and his bride on their wedding night. A chivaree was a tradition that the kids of Stansbury loved because we made money from it.

The day Jack brought his new wife home after the wedding ceremony, families brought food, beer, and accordions and gathered to celebrate in Jack's front yard. As soon as it was dark, the adults quit dancing and went home to let Jack be with his new wife.

But we kids ran for our supplies, then sneaked close to their darkened bedroom window and beat on tin cans, old tubs, buckets, or anything else that made a lot of noise. This noise was deafening and didn't stop until Jack opened the bedroom window and tossed out coins, mostly nickels and dimes. We "serenaders" scampered to find the coins in the dirt, in the dark, and then satisfied with our loot, went home. How we loved doing this!

○ ✕ ○ ✕ ○

The summer I was twelve, Jimmy ten and Johnny eight, Momma's wealthy sister Ann asked if we kids could spend two weeks with her and Uncle Zeke in Salt Lake. Johnny didn't want to leave Momma, even for two weeks, but Jimmy and I could hardly wait. So Aunt Ann sent money for two Greyhound bus tickets.

When the day came for us to leave, Momma and Daddy took Jimmy and me to the bus depot in Rock Springs. Up until this time, we had never been away from our parents, let alone traveled by ourselves. We were both excited yet somewhat apprehensive as we waved goodbye when the bus pulled away from the depot. Jimmy and I sat in seats next to each other and didn't talk to anyone else the whole trip. We found ourselves glued to the windows as we rode through Wyoming and into Utah. The minute we crossed the state line, Jimmy turned to me and said, "Marilyn, look how green everything is in Utah, and I don't see tumbleweeds blowing anywhere. This is way different from Stansbury." I agreed we were entering new territory.

Aunt Ann and Uncle Zeke were waiting for us when the bus pulled into the Salt Lake depot. It was a good thing because Jimmy and I started being scared the moment the bus traveled down Parley's Canyon, and we saw ahead of us the big city of Salt Lake. "If we don't like it here, we can go home whenever we want, right, Marilyn?" Jimmy asked.

"Don't worry. All we have to do is call Momma and Daddy if we want to go home, and they'll come get us. But, don't be afraid; we're only going to stay here for two weeks," I assured him.

Aunt Ann and Uncle Zeke ran to us the moment we got off the bus. After hugs and kisses, Uncle Zeke loaded our two suitcases in their beautiful baby blue Cadillac. Jimmy whispered in my ear, "They have a new Cadillac, Marilyn." No one in Stansbury owned a Cadillac. We'd always heard that a Cadillac was a car only rich people could afford.

He and I sat quietly in the back seat of the car taking in all the beauty of Salt Lake as we answered Aunt Ann's questions about our trip and everyone back home. We couldn't get over how green everything was, how big the trees were, and the abundance of grass and flowers. Big storefronts lined the streets along the way, and restaurants were everywhere.

Uncle Zeke and Aunt Ann Snyder at their dining room table in Salt Lake City. They opened a whole new world for us Nesbit kids.

When we turned onto a residential street, we saw beautiful, big houses made of brick, so unlike the small wood framed UP houses. The yards looked like those I had seen only in magazines.

"Well, this is it, kids," Aunt Ann said as we pulled into their driveway. "Let's hurry inside so I can show you around and help you get your suitcases unpacked. I've got your room all ready for you."

Jimmy and I hesitated at the doorway. Even though we remembered to remove our shoes when we entered the house, as Momma had instructed, we hesitated to walk on the thick wall-to-wall cream-colored carpeting for fear our stocking feet would leave footprints. Aunt Ann opened the drapes simply by pulling a cord on the right side of each drapery, revealing big picture windows. The fragrance of fresh flowers from an arrangement in a crystal vase above the fireplace drifted in the air. We were afraid to touch anything, but when Aunt Ann noticed our hesitancy, she put her arms around our shoulders and led us through the house, "Now come on. You can't hurt anything. Your feet aren't going to hurt the carpeting. Act as if you were in your own house. You can't imagine how happy we are that you're here. Uncle Zeke and I don't have any children, so we'll treat you like you're our own." Her words put us more at ease. "The room you'll be staying in is right down the hall, next to the TV room. Come on, I'll show you," she said. We followed Uncle Zeke who was carrying our bags.

When we passed the TV room she explained that television was something new, like having a picture show right in your own home, where we could watch many of the same shows we listened to on the radio. Jimmy and I could hardly wait to see what television was like.

The room Jimmy and I were to share had twin beds with a night stand in between and a big picture window overlooking the picturesque front yard. A delicate ballerina figurine stood in the center of the mirrored dresser and porcelain boxes were scattered across the top. Uncle Zeke laid our suitcases on each of our beds, and we began putting our things in the dresser drawers and closet that was big enough for our whole family to hang their clothes in.

Aunt Ann showed us the bathroom which gleamed fresh and white from the carpeting to the tiled walls. A silk shower curtain hung in swags beside the bathtub. A mirror extended across the top of the vanity and reflected back the lights above it. The fragrance of the bar soap set in decorative trays near the sink filled the room. A glass-shaped decanter held colorful bath oil beads in the corner of the vanity.

Uncle Zeke took us back into the television room to give us a preview of TV. When he turned on the set, our mouths dropped open when we got our first glimpse of the shows on the black and white television. Uncle Zeke showed us how to operate all the buttons, adjust the rabbit ears that brought the picture in clearer, and said we could watch it anytime we wanted.

"Does everyone in Salt Lake have one of these?" I asked.

"Yes. And soon you'll be able to get television where you live too. It's something brand new. I understand they will be coming out with colored television soon. But, right now, why don't you wash up for dinner?"

"Aunt Ann, which towels are we supposed to use?" I asked.

"Just use the towels on the rack to the right of the tub."

"But we'll get your white towels dirty. Don't you have some dark-colored towels for kids to use?"

"Oh, Marilyn don't worry about the towels. If you lather your hands really good with soap, there won't be any dirt left to wipe onto the towels," she replied with a smile.

Jimmy and I headed for the bathroom. We washed our hands with soap and then stopped and looked at each other while our hands were still in the sink. "Let's just wipe our hands on toilet paper this time," I instructed.

"Yeah. That's what I was thinking," Jimmy replied. He cupped his wet hands together and reached for the toilet paper, pulling off a long piece from the dispenser. I did the same.

The corner windows of the dining room sparkled in the setting sun. Flowered wallpaper covered the walls. A white lace tablecloth draped the table which was perfectly centered below the chandelier. Jimmy and I were unsure how to act. No matter how Aunt Ann and Uncle Zeke tried to make us feel at home, we felt poor and out of place. We were relieved to make it through the meal without spilling anything.

After dinner I helped Aunt Ann do the dishes while Uncle Zeke took Jimmy to the TV room. I could hardly wait until the last pan was put away so Aunt Ann and I could join them. That evening, I saw what was to become my favorite program: *The Ed Sullivan Show* with all his celebrity guests. Up until that moment, I had never even heard of Ed Sullivan but I was mesmerized.

Later that night when all the lights were out, Jimmy and I stretched out on our beds silently reliving the day. Breaking the silence, I said, "Kind of strange not to hear the mine whistle, isn't it? Everything is so quiet here. I keep listening for the sounds of the mine coming from below our beds."

I thought of Momma and Daddy's voices on the phone when we'd called home to let everyone know we'd arrived safely. "I miss Momma and Daddy and Johnny. I wish we could all live here," I said as I rolled over and pulled up the covers.

○ ✕ ○ ✕ ○

The next two weeks were filled with experiences we'd never had before. One day Uncle Zeke drove us up through one of the canyons in the mountains above Salt Lake, where Aunt Ann bought fresh fruits and vegetables at one of the roadside fruit stands. Jimmy and I looked at each other with eyes as big as golf balls when she bought each of us a flat of raspberries. We could hardly wait to get back to the house to eat a big bowl of raspberries and cream.

We spent our evenings watching television until we fell asleep in front of the screen. We went to the Salt Lake Zoo; Salt Aire, where you float in the salty water, even if you don't know how to swim; Grand Central, the first discount store we'd ever seen; and Auerbachs, the Bon Marche, and ZCMI department stores. The biggest department store that we had ever seen before this trip was the UP Store in Rock Springs.

Everywhere we went, Aunt Ann and Uncle Zeke bought us everything we liked and more. And they always remembered to buy for our little brother back home. Jimmy and I were constantly amazed as Aunt Ann paid for the things she bought us. It was as if we had stepped into a whole new world.

Off and on, kids from the neighborhood came over to invite us out to play. One family had a swimming pool in their backyard, which was usually filled with neighborhood kids and soon included Jimmy and me.

On Sunday everyone in the neighborhood spent practically the whole day in church, the Mormon Church. Aunt Ann and Uncle Zeke were the only ones in the neighborhood who were not Mormons: Aunt Ann was Catholic and Uncle Zeke was Jewish, though I never remember him going to church. Even though Jimmy and I were children, we still felt a little different and left out. When I asked Aunt Ann about it, she went into a deep explanation of how things were in Utah. I realized then how Uncle Zeke must have to struggle to have a successful business in Salt Lake. Religion became something we tried not to talk about when our new friends asked, "How come you aren't Mormon? Ya know, you won't go to heaven if you aren't Mormon."

"No sir, that's not true, huh, Marilyn?" Jimmy would snap back and then everyone would go silent.

The Saturday before we were to go back home, Uncle Zeke had one more surprise for us. He took us down to the wholesale magazine company he owned and let us pick out as many magazines and comic books as we wanted. Warehouse men brought over empty boxes for us to pack them in. I picked out issues of *Seventeen* magazine, *Your Hit Parade,* movie magazines, and comics. Jimmy concentrated strictly on comics. When we finished, we had five boxes filled to the brim. After that, we had a picnic lunch at Lagoon, a popular amusement park.

The day we were to leave, Aunt Ann prepared eggs and bacon for breakfast. After everyone was seated at the table, Jimmy asked, "Do you have any ketchup, Aunt Ann?"

"Ketchup?" She looked at him curiously.

"Yes," Jimmy persisted, "to put on our eggs."

Uncle Zeke threw his head back and laughed, "Look, Ann, these coal miner kids actually put ketchup on their eggs! I've never heard of it!"

"Try it," I said. "Eggs taste good with ketchup."

After breakfast, we helped Uncle Zeke load our luggage and boxes of magazines and comic books into the car. We loved being in Salt Lake, but we missed Momma, Daddy, Johnny, and our friends back home. We were also looking forward to our family vacation at Granite Hot Springs, a campground near the mountains south of Jackson Hole.

As we boarded the bus to head home, we gave our aunt and uncle big hugs and kisses. Tears welled up in Aunt Ann's eyes. "Don't worry, Aunt Ann," I said. "We'll be back. Thank you for being so nice to us and buying us all those things and the comic books."

"Yeah," agreed Jimmy, looking at the comic books he was holding under his arm.

The bus driver pulled the door shut, and Jimmy and I waved until we couldn't see them anymore. "Are you sad we're leaving, Marilyn?" Jimmy asked.

I turned my head toward the window so he wouldn't see the tears in my eyes, "We couldn't stay forever," I replied. "Momma and Daddy would miss us too much. But they sure are nice. Maybe someday Daddy will move us down here."

○ X ○ X ○

When Momma, Daddy, and Johnny spotted the bus from Salt Lake, they began waving their hands in the air. Jimmy and I could hardly wait for the bus to come to a stop so we could get off. People in the seats behind us quickly filled the aisle leading out of the bus. "Come on, Marilyn. We better cut in the line, or we'll be the last ones off," Jimmy instructed.

Suddenly we could not wait to get off the bus and into the arms of our parents. I remained seated out of the decorum my mother had taught me, but Jimmy stood up and when he saw an opening in the line, he yanked on my arm, pulling me into the moving line of passengers. When he did this, a part of the full skirt of my dress caught on a sharp edge of the seat ahead of me, and I tripped, falling into the aisle. My entire skirt ripped off at the waist. People looked down at me curiously for a moment, then stepped around me and continued to clamor toward the door. I pulled myself up, gathered my skirt and held it at the waist, and then slapped Jimmy hard on his back. He'd not only ruined my skirt, but he'd embarrassed me in front of everyone on the bus.

Suddenly it was too much and I was crying. I held my skirt around me and ran off the bus into Momma's arms. She and Daddy, along with everyone on the platform, were laughing so hard. "It isn't funny, Momma." I wanted to kill Jimmy. Thank goodness the back zipper held my dress on me.

"Oh, Marilyn, it was an accident. We can fix your dress when we get home," Momma consoled. But it wasn't the greeting I'd longed for.

On our drive out to Stansbury, I couldn't help but notice the stark contrast between Utah and Wyoming. The two weeks in Utah opened my eyes to a lot of things. Everywhere I looked, things around us were brown and dried up, even the creek along the highway was dry. Salt Lake had been so green. As we got closer to Stansbury, the wind came up, stirring up the sandy plains and rolling tumbleweeds. It reminded me of scenes from a movie I had seen, *The Grapes of Wrath*. A feeling of sadness came over me. I'd made a fool of myself at the bus depot, Momma and Daddy laughed at my clumsiness, and the whole country-side was ugly and desolate. I lowered my head and looked down at the floorboard of the car, forcing myself not to cry. I didn't want to explain my tears to my parents.

Momma seemed to enjoy cooking on a Granite campstove.

Momma broke into my thoughts, asking me if it felt good to be home, and even though I said yes, at that moment I didn't like being home. Jimmy chattered all the way home about our trip. "Wait until you guys see all the things they got for us—you too, Johnny. Everything they got for me, they got for you too. We'll probably never have to buy comic books again, huh, Marilyn? We brought boxes of them home with us, and Uncle Zeke is going to ship more to us every single month. And, Momma, we didn't have to pay anything for 'em. There were zillions in his warehouse!"

Finally Daddy interrupted, "Well kids, I'm glad to hear you had a good time, but soon as we get home, we have to pack up so we can go up to Granite tomorrow."

○ ✕ ○ ✕ ○

*Granite Hot Springs, with its warm pool, became a favorite vacation site
for the coal mining families starting in the 1930s.*

Each year, Daddy got two weeks vacation. Usually we went to Granite
Hot Springs, a camping area about thirty miles north of Pinedale,
Wyoming, in beautiful Hoback Canyon. Other mining families went
there too, mainly because it was inexpensive, it was close, and everyone
loved this mountain area. Granite offered something for everyone—
camping, fishing, places to explore, and swimming in the outdoor naturally-heated mineral pool anytime of the day or night, at no charge.
What made this even more fun was that many of our friends from
Stansbury and the other mining camps were there, too.

Our first stop along the way to Granite was the Eden-Farson General
Store where we each got one of the homemade ice-cream cones that
were their specialty. Tippy remembered the slow moving river that ran
near the store and plunged in as soon as we let him out of the car. When
we were ready to leave, Daddy whistled and Tippy swam over to the
shore and shook the water from his fur before we rubbed him down
with towels. As a special treat, Daddy bought Tippy a small cone.

Then we headed into the north country, toward the majestic mountains once traveled by Jim Bridger. From the moment we left the last
town, Pinedale, our eyes were glued on the beauty as we went deeper

and deeper into the Hoback until we reached the Granite turn-off.

"There it is, Daddy!" Johnny hollered in excitement when he spotted fishermen in the rushing Green River that was near the Granite Hot Springs marker. Daddy turned off the main highway onto the gravel road that ran along the river leading to Granite. The smell of the fresh mountain air told us we were almost there.

We entered the campground and slowly drove along the winding dirt road nestled under huge pine trees, looking for a place to pitch our tent. Suddenly I sang out, "There's a campsite, Daddy, and it's real close to the outhouse." He pulled the car over, and we kids jumped out.

My brothers and I loved the big, canvas tent that Daddy had bought at an army surplus store. We took plenty of lanterns to set up around our campground or to carry with us when we went for a swim in the pool after dark. When we did swim at night, everyone set their lanterns around the pool so the whole area was well lit, and the water glistened with reflected light.

The campground was well kept by the Forest Service. Every day, forest rangers stocked each campsite with foot-long logs cut from fallen trees. All the campers had to do was split logs for kindling. While my brothers watched Daddy and a friend set up the tent and unload all the gear, I helped Momma with the food. "Marilyn, you and your brothers take these melons, pop, and your father's beer down to the creek and put them in the water near the bank to keep nice and cold," Momma instructed, "while I get the campfire going so I can cook supper."

Later we'd go to the swimming pool about a mile from the campground, up a gravel road that trailed alongside the rushing Green River. Hot mineral water spilled from a high mountainous rock formation into a round, cemented pool below. A foot-wide cement outer edge ran all around the pool where swimmers sat and where we placed lanterns for night swimming. We all loved to cannon ball off the wooden diving board at the deeper part of the pool.

I learned to swim and dive at Granite. I was nine years old and kept begging Daddy to teach me to swim. "How are you going to teach me?" I kept asking as he led me toward the deep side of the pool and out onto the diving board. "The same way my father taught me," he said as he picked me up and threw me into the middle of the pool. When I hit the water, I heard the rush of the water in my ears as I plunged deep down into the water, then suddenly I thrust upward to the surface while I thrashed my arms and legs in panic. I knew I was going to die and recalled Momma saying a person drowns after going under the water three times. I knew I was drowning. I kept kicking my legs and pulling at the water with my hands until suddenly I felt the edge of the pool. I came up and grabbed on for dear life. Water spurted out of my nose and mouth. Daddy stood on the ground right above me and laughed as he reached his arms out to give me a hand.

"That wasn't funny!" I yelled. "I could have drowned, ya know."

Calmly he said, "I had my eyes on you every minute, Marilyn. Look over to the diving board. That is how far you swam, all by yourself." From that moment on, I spent all my time in the deep part of the pool, swimming on my own.

Late one night Daddy and Momma thought we kids were asleep and sat around the campfire talking. I overheard Daddy say, "Margaret, I sure wish we could afford to take the kids somewhere else for vacation once in a while. I wonder if they get tired of coming here, year after year?"

I wanted to pop out of my sleeping bag and tell him all the reasons we loved every moment at Granite: waking up each morning to the smell of coffee brewing over the open campfire as Momma fried bacon and eggs in her cast iron skillet; hiking all over the mountains with Tippy; hauling spring drinking water from the campground faucet and carrying buckets of water from the river to use throughout the day; following the trail down to the river to fetch the ice-cold melons and bottles of pop and beer we stored there; fishing with Daddy and the men and helping to clean the catch of the day with our sharp pocketknives so Momma could fry them that night for supper; joining the adults for a game of horseshoes or darts; sleeping on cots in our army tent under warm quilts and blankets; sometimes falling asleep to the soft sound of raindrops on the roof of the tent; feasting on Momma's meals, especially when we had fresh garlic sausages, wrapped in foil, and baked in the hot ashes of the campfire; walking back from the pool with only lanterns to show the way; roasting wieners and marshmallows over the open fire;

My brothers Jimmy and Johnny loved camping in our tent at Granite Hot Springs in the summers.

telling ghost stories; listening to the men talk about the mine or wartime as they played poker; seeing the loving glance our parents gave each other when we sang certain songs around the campfire. All this without the distraction of radios, phones, or call-outs from the mine. Granite Hot Springs was the best possible place to spend a vacation.

Right before I fell asleep that night, I just had to let Daddy know how much I loved it all. I got out of my sleeping bag and stuck my head out of the tent in the quiet of the night just long enough to softly say, "Daddy! This is my favorite place in the whole wide world to go for vacation. Thanks so much for bringing us here again this year!"

Women started working in quality control in the mine during World War II when they were allowed into non-traditional jobs because of the worker shortage. Here, in 1946, women pick foreign matter off the conveyor belt as the coal passes through the tipple in Stansbury. (Department of Interior, Russell Lee)

CHAPTER ELEVEN
PENDULUM SWINGS

I N JULY 1952, while Jimmy and I were in Salt Lake City for our second summer, UP closed the Winton Mine, the oldest mine in the region. As a twelve-year-old, I was not seriously affected by the closing of this mine. The booming town of Stansbury was all I cared about. But, as time went on, even I saw the impact as the Winton store, clinic, pool hall, post office, school, and the community hall closed and the buildings were sold and moved from their foundations. Watching the demise of the town was heartbreaking. South Pass City was the only ghost town I had ever seen, and now I was seeing with my own eyes the death of Winton and the birth of a ghost town.

One hundred seventy-seven men lost their jobs at the Winton mine. Most were older, but not yet eligible for retirement. Fortunately, most of the miners were able to transfer to other mines in the area. Those who loved their homes chose to stay in Winton as long as they could and commuted to the mines in Stansbury, Reliance, Superior, or Rock Springs.

○ × ○ × ○

On a crisp, clear summer day in 1953, right before I entered the seventh grade, I got the biggest surprise of my first thirteen years. Daddy announced that it was time I had my very own bedroom. He'd build a room in the basement for my brothers to share. Up until this time, since all company houses had just two bedrooms, my brothers and I slept in the same room with a double bed that they shared and a single bed for me.

After supper one evening, Daddy and Momma sat down at the kitchen table, and we kids hovered around as he began sketching the plans for the downstairs bedroom. My brothers, who were eager to assist, gave input on what their bedroom should look like. Before long one corner of the basement was taken up by lumber and sheetrock.

We kids had entertained ourselves, summer and winter, in the basement. With the new room taking up one whole corner, we would no longer be able to have all of our friends come over to roller skate around the coal furnace while listening to music played on our 45 record player, nor would we be able to hold Halloween carnivals there. But this was a sacrifice we were willing to make as my brothers and I really wanted our own bedrooms.

When the bedroom project was completed, Momma bought the boys a new twin bed set at the UP Store. She gave the old metal bed they had slept in for years to a family with many children. The day after my brothers moved downstairs, Daddy took me aside and told me that he would buy me a brand-new bedroom suite. My mouth dropped open in utter amazement for I had never known any girl in camp to get a new bedroom set of her own. "You sure you can afford this, Daddy?" I asked.

He explained that a railroad freight car transporting furniture through Rock Springs had been damaged in a derailment. The UP announced to local miners that anyone interested in purchasing the damaged freight at a discount could examine it. So Momma took me to Rock Springs to look.

At the freight depot, UP yardmen were unloading the damaged freight into a warehouse. I immediately spotted the furniture I wanted. "Oh, look, Momma, look at that set over there with the big beveled mirror," I shrieked, running toward it. I could feel my heart beating fast, and I prayed my parents would have enough money to buy it.

Momma said, "Now don't get so excited, Marilyn. We'll have to check it over really good first and then find out how much they want for it." I waited for what seemed like hours as Momma talked with the freight manager. Then she said, "I'll call your father and, if he approves, he'll make arrangements to have it delivered." She was on the phone to Daddy only a few minutes before she hung up and looked at me with a smile on her face. "Looks like it's yours, Marilyn." I ran back over to the furniture and rubbed my hands across the smooth piano finish of the dresser top.

Except for a small scratch on the back of the dresser—and who would see that?—it was in perfect condition. Made of the blonde wood which was so popular in the Fifties, the dresser had several drawers in the middle, plus a hinged door on each side with drawers behind it. The hidden drawers had fancy shaped glass inserts so that you could see inside them. The large matching mirror that was to hang above the dresser had a beautiful six-inch fluted mirror casing with a beveled edge. For many years afterwards when I looked at that furniture I felt loved.

I asked my parents if I could finally decorate my room any way I wanted. I'd made a scrapbook of various bedroom pictures I'd cut from magazines, and I knew exactly how I wanted mine.

"Okay, okay Margaret," Daddy repeated to Momma. "Let her fix up her room exactly how she wants it. But, for God's sakes, don't put us in the poor house doing it."

One day while Daddy was at work, Momma painted my bedroom ceiling and three walls a chocolate brown, and on the weekend, Daddy hung wallpaper on the remaining wall. I had picked out a pattern illustrating a quaint village scene. I sat on the floor and watched Daddy cut and paste each strip to create the village design, and even though I knew he hated handling wallpaper, he never once complained.

Daddy said Momma and I could order, from the Spiegel's catalogue, sheer Priscilla curtains, chenille throw rugs, and a matching bedspread, all in my favorite color—pink. I couldn't wait to show my friends. My bedroom soon became the talk of the camp. I overheard one lady commenting to a friend as she left our house after Momma showed her my room, "My land! Can you imagine fixing up a young girl's room fancier than her parents'? And, wallpaper, no less. Where does she think she lives, in Hollywood?" Momma said to just ignore their comments.

Daddy had to carry the big box from Spiegel's home from the post office. It was just too big and heavy for my brothers and me to maneuver. The minute he set it down in the middle of the living room floor, I ran to the kitchen and grabbed a kitchen knife to cut it open. Everything Momma ordered was in this one box. The contents were a vision of pink, from the sheer curtains to the fluffy, soft chenille rugs.

At supper that night, I told Momma I wanted to cash in one of my government stamp books at the post office so I'd have some spending money the next time we went into town. I also wanted to sell my balls of aluminum foil at Doan's Hide and Fur.

As I headed toward my bedroom that night, I heard Momma say to Daddy, "John, tonight on the news was the second time today I heard them mention something about a new disease called polio. No one seems to know what causes it or how to treat it. I'd better give Dr. Muir a call in the morning."

○ × ○ × ○.

On July 22, 1953, the first five cases of polio were reported in Rock Springs. On August 11, 1953, three more cases of polio were reported. Everyone was suddenly alert, and parents in the coal camps were scared to death about the mysterious disease, especially when they heard of the devastating and long-lasting effects it had.

CHAPTER TWELVE
MY WORLD ENLARGES

ARTICLES CONTINUED TO appear in the Rock Springs newspaper regarding the ongoing closure of Winton. Often they were only a paragraph or two long and without a headline. But for those living in the other mining camps, it was enough to remind us almost daily that it could happen to us too.

I dreaded getting up the morning I entered junior high. For the first time I wouldn't walk to the elementary school with my brothers, but instead I'd take the bus to Reliance School to begin the seventh grade.

I hugged Momma tightly and whined, "I'll hate that big school, Momma. I won't even be able to come home for lunch anymore!"

"Oh, come on now, Doll," Momma coaxed, squeezing me close to her. "You're going to love the seventh grade. You'll have so many things to tell me and your dad tonight. But, you'd better hurry; I hear your friends out back. Here's your lunch card. It's good for one month. The cooks won't let you eat until they punch it, so don't lose it."

I carefully tucked the card into my purse and then headed out to walk with my friends down to the UP Store where the Reliance bus picked up students. I took one look back at Momma standing on the back porch steps watching me and heard her yell, "I love you, Doll. See you when you get home."

○ ✕ ○ ✕ ○

Kids stood at the side of the store waiting for the bus to arrive. Off in the distance I could see an old yellow bus wobbling toward us on the winding highway. It stopped in front of the store, the driver pulled the handle opening the bus door, and we climbed aboard. The kids talked and joked with each other during the three-mile drive to Reliance, and I wondered if they were as nervous as me.

The Reliance school was so much bigger than the elementary school in Stansbury. Older students, many of whom I did not recognize, milled around the halls. Feeling bewildered, I shuffled through my papers looking for my class schedule. "It's confusing, isn't it, Marilyn?" my friend Carolyn whispered as we compared class schedules. "Gosh, yes," I replied. "Looks like we have most of the same classes, so let's just stick together."

Ira Russell, the school superintendent, made an announcement over the loudspeaker directing all incoming students to a room for orientation. Carolyn and I bounced off other students through the halls until we found the room and took seats next to each other. Mr. Russell welcomed us and went over the rules. He gave us a tour of the building and then it was time to go to our first class. The morning passed quickly, and soon it was time for lunch.

My final class that afternoon was music with Mr. Art Nyquist, the band instructor, who sat at a desk heaped with sheet music. A tall, stately man about Daddy's age, he wore a suit, crisp white shirt, and tie. A lock of blond hair hung across his forehead just above his glasses that always rode down on his nose. The moment he started speaking, I knew I was going to learn a lot from this man. I'd always loved music and wanted to learn everything, whether it was playing musical instruments or singing in the chorus. I signed up for both band and chorus. But I had to decide which instrument to play. The school supplied instruments, some brand new, and charged a minimal rental fee. I told Mr. Nyquist I'd have to check with my parents about the rental fee before I decided on an instrument.

By the end of the day I had met students who would become my friends—like Eva Page and her older sister Priscilla, Ronnie Welsh, Jimmy Burns, John Sawick, Billy Thomas and his sister Carolyn, Leone Theinpont, Linda Harris, Virginia Agar, Carl Pearson, Judy Gaylord, and Carmen Esparza.

The day I had dreaded so much, I found myself loving.

That evening over the dinner table, I could hardly contain myself long enough to listen to my brothers report on their first day back at

In Stansbury, waiting for the school bus to Reliance: Nora Lee Hereford, Sandra Seppe, Geraldine Guigli, Carolyn Flipovich, Veronica Bozner, and me. The mine office is in the background.

school. My whole family listened intently as I told them about the new friends, the instructors and classes, and how good the lunchroom food was. "Oh, before I forget, I have to rent an instrument for band. I have the rental price list Mr. Nyquist gave me in my notebook. Oh, Momma, I hope we can afford the rental fee."

"Don't fret about the fee," Daddy replied. "But what instrument do you have in mind?"

When I hesitated he said, "Why don't you give the clarinet a try? The clarinet always carries the melody." So I agreed.

I stopped eating for a moment, looked up at Daddy, and blurted, "Daddy, I want to go to college."

"College?" Daddy asked in amazement. "How'd we get from your first day of junior high to college?" Looking at me from across the table, he said, "Well, I'll make you a deal. If you go through high school and graduate as the valedictorian or salutatorian, I'll help you go to college. But, you know what that means. You'll have to make the best grades in your class starting this very minute until the day you walk off that graduation stage. Think you can do that for six years, 'cause I won't pay for play?"

"I know I can do that, Daddy! Just you wait and see. I want to be a teacher," I replied assuredly.

"Then work hard toward that dream, Marilyn, and get yourself a good job anywhere but in a mining camp," he encouraged.

○ × ○ × ○

Seventh grade was truly the turning point in my life. I soon realized that junior high was going to be a lot more competitive, but competition never threatened me. In fact, I thrived on it. My parents paid the rental fee for a clarinet as well as the piano lessons Jimmy and I took from Mrs. Nyquist, the band director's wife. Our piano lessons were one dollar each a week. Since the Nyquists lived just down the street from where I lived in Stansbury, they seemed to take a special interest in me.

My teachers included the beautiful Jean Peppinger for math, my favorite subject; Stephen Owens for science; Robert Armstrong for social studies; and Chez Haehl for English. Marjorie Russell, my home economics teacher, taught me to be an excellent seamstress until I could eventually make all my own clothes. I was in awe of her charm and hoped it would rub off on me. Her make-up and hair were perfectly done, and she always wore flawless fingernail polish on nails shaped like rose petals. From the day I met her, I strived to be just like her.

Our first sewing project sounded easy enough—make an apron out of white muslin to wear during our cooking classes. She showed us how to put the apron together, insisting we always sew three rows of stitches to pull for gathering. I was so eager to finish that I hurriedly moved on to the next step after only two rows of gathering stitches, thinking she'd never know the difference.

While we were sewing, she walked around the classroom observing everyone's work. She stopped by my sewing machine, picked up my apron, and examined it carefully. I held my breath, hoping she didn't notice my shortcut. She snipped a thread in the middle of each row of gathers and pulled the material taut. Out came all my gathers. My eyes bulged out of their sockets when I saw what she had done, and I didn't know what would happen next. I almost cried.

"If you want to do it right, three rows! Start all over and put the gathers back in." She laid my apron over my machine. From that day on, each time I put gathers in any garment I relive the moment and feel my face blush.

I learned to play not only the clarinet, but the oboe and French horn as well. I also progressed from Brownie Scout to Girl Scout. Our leader, Helen Henderson, organized cakewalks and bake sales so our troop members could pay dues without asking our parents for money. I continued to be active in 4-H under the direction of Lola Nielson's elderly mother, Minnie Shultz.

○ X ○ X ○

Our house was always filled with kids, especially after Momma became a Cub Scout leader and Daddy hung a punching bag in our basement to teach boys how to box.

One evening Daddy announced that I was old enough to begin first aid training so that I could participate in the yearly contests sponsored by Union Pacific. This year he was in charge of organizing the Stansbury teenage and adult men's teams. The kids' team members had to be at least twelve years old and a Girl or Boy Scout. Each team had six members—a captain, a patient, and four assistants—in specific age groups.

The students on my team were Carolyn Pecolar, Darlene Fabiny, Sherrie Jenkins, Carrie Palcher, and Sharon Logan. Usually the smallest girl was the patient, as she would be the easiest to lift. Sharon Logan was our patient. Daddy chose me to be our team captain.

Two or three nights a week, all the Stansbury teams met to practice at the Stansbury Community Hall. Each team was assigned a six-foot long, white, wooden footlocker built by the miners. Daddy went early to practice so he could retrieve the lockers from the back room before everyone arrived. Each locker contained wooden boards specifically cut to make splints and a stretcher, plus blankets and cravats. Cravats were rectangular-shaped muslin cloths which we folded to make ties for splints and stretchers. We padlocked the lockers at the end of each practice.

The first session we learned basic first aid procedures such as how to fold cravats and tie a perfect square knot. We all struggled to make perfect square knots. Then each team member, teenagers and adults alike, received the 330-page Bureau of Mines' *Manual of First Aid Instruction*. Daddy insisted we read each assigned lesson from the manual before practice. If we didn't, we had to go to sit alone to read the lesson while the others practiced. This was so embarrassing that very few forgot to read their lesson. We learned how to identify and treat injuries including giving artificial respiration and how to transport patients properly.

Then we put our skills to timed practice tests. Our instructor handed the team captain a typewritten account of an accident, the location, the environment, the condition of the patient, and time constraints. The captain, who stood at the foot of the patient, read the accident details aloud to the team members, and together they determined how to administer first aid to stabilize the patient until medical help arrived.

Each member of the team had to attend all meetings and to study and work hard to achieve the goals: to be able to apply first aid in a real life situation should it arise, but also — seemingly more important to us kids — to do well at the annual first aid contest the following summer and win prizes ranging from money to radios, Kodak cameras, or a complete set of Samsonite luggage with a name plate. Teams from all the mining towns participated and the competition was fierce. For a team to take first place was a big honor for their whole camp.

○ ✕ ○ ✕ ○

During the winter, most of the kids' activities were inside except for sledding on the South Camp hill or ice skating in our front yard. Daddy flooded our entire front yard with the garden hose, day after day, until the surface was hard and smooth. He hung floodlights on the outside of the house, lighting the whole area. This drew kids from all over camp, and on those nights Momma made hot cocoa with marshmallows and asked everyone inside.

We also used Momma's old cookie sheets or pieces of cardboard boxes for sledding. Some kids used large coal shovels for sleds, and everyone shared whatever they had. After several years, we got two sleds for Christmas. We rubbed a bar of soap or piece of paraffin wax up and

We kids loved to go sled riding on the South Camp hill. Back: Ronnie Henderson, me, Bobby Henderson, my brother Jimmy Nesbit. Front: Michael Corney, Jerry Hereford, and my brother Johnny Nesbit.

down both sled runners to make them go real fast. Sometimes on the weekends when everyone was sledding during the day, parents hooked a long rope to the back bumper of a vehicle and pulled a sled with riders along the snow-covered camp streets.

On more than one occasion, Daddy was called back to the mine in the evening. He sometimes went to another mine portal above camp, to check on his men who were about to finish a project and come up out of the mine. My brothers and I begged to go with him. It was exciting to stand beside Daddy at this portal and listen to him talk by radio to the men who were miles underground. In the black of night he would holler out, "Everyone aboard?" A faint, distant reply could be heard, "Everyone accounted for, Johnny. Let 'er go!" whereupon Daddy threw the switch to set the mantrip in motion. We were so proud of our dad and the respect he received from people in the camp.

Momma's homemakers' club, the Busy Bee Homemakers, sponsored by the State Extension Office and under the watchful eye of the Home Demonstration Agent Alma Shelt, helped ward off the long winters with more meetings. Daddy playfully called it the "Home Wreckers'" club and Momma just rolled her eyes at him.

We were all thrilled the day a big box came through the mail simply addressed to "The Nesbit Kids." The return address was Ann and Zeke Snyder—Aunt Ann and Uncle Zeke—who had promised to send us more out-of-date comic books and magazines. The box was so heavy we had to take turns carrying it on the way home from the post office. Our friends tagged along on the trip, making a little parade, and we dumped the contents in the middle of the living room floor where everyone sat for hours looking through all the books and magazines.

While kids were busy reading all the comics and magazines, some neighbors came over to visit with Momma and Daddy. We knew it had to be about something serious when Momma told us to stay in the front room. Even though they spoke in low tones at the kitchen table, I could hear bits and pieces of what they said especially when the lady tearfully exclaimed, "Oh, my God. What will people think when this gets out?"

I learned later that their young daughter was pregnant. In those days, if a girl became pregnant, she was forced (or allowed) by parents to marry the baby's father. The only other choice was for her parents to send her to live with relatives or to a convent until the baby was born. A family decision was made about whether to keep the baby or give it up for adoption. The pregnant girl was labeled wicked and wild. Many times, when the girl's father found out, he beat her severely with a belt.

I could tell Daddy was upset by the situation, though he never mentioned it to me. Then, one night I was sitting with him in the yard when out of the blue he turned to me and said with an angry look on his face, "Don't you ever bring shame on the family by getting yourself in the family way before you are married 'cause if you do, I'll kill the son of a bitch who did it to you and then kill myself."

"Daddy!" I shouted back. "Why would you say something terrible like that? It scares me to hear you say that. That's crazy, and I'm going to tell Momma right now what you just said," I sobbed as I ran into the house into her arms. "My lands, Girl, what is all this commotion about?" Gasping for breath between tears, I told her what Daddy had said.

"Now calm yourself down while I go and talk with your father." I ran into my room, slammed my bedroom door behind me, and flung myself across my bed crying.

Soon Daddy came into my room and sat at the foot of my bed. "Marilyn, I shouldn't have said that. Your mother is really upset with me. You have to understand where I was coming from. I feel terrible about what happened to that girl and her family. It would kill me if that happened to you, and I will do everything in my power to prevent that from ever happening. My words were harsh but . . . ."

"Come on, Marilyn," Momma walked in and interrupted, taking me by the hand and leading me into the living room. "Your favorite show is coming on the radio right after the news."

I sat quietly with my brothers in front of the radio, Momma sat in her favorite chair and picked up a piece of crochet work, then Daddy came in and sat in his chair. "Well, one thing we can be glad of," Momma began, "the news doesn't seem to mention polio in the winter like it did in the summer. It will be a miracle if they find out what causes that horrible disease and come up with some kind of cure."

"Okay, everyone be quiet," my brother Johnny urged. "*Sergeant Preston* is coming on."

○ x ○ x ○

During that first year of junior high school, I experienced something that was completely new: I met a boy from Rock Springs at a Saturday matinee who was to become my first boyfriend. Joe was in the ninth grade, two years older than me. After that first meeting, we talked on the phone and met every Saturday afternoon at the Rialto or at youth activities held at the Mormon Church. I knew Daddy would never approve of me having any boyfriend, so I kept it from him.

One cold, wintery night Daddy was called back to the mine after supper. As he headed out the door he called back to Momma, "I don't know what time I'll be back. My men have big problems down in the mine." After he left, I asked Momma if I could use the phone.

I called Joe and after we talked for a while he blurted out, "Ah heck, I'm going to ride my motor scooter out to see you. My parents won't be home until late 'cause this is their bowling night."

"But, it's snowing and freezing outside," I warned. He assured me he'd dress warmly, and it would take only fifteen minutes to drive out. I couldn't deter him.

Since I couldn't ask him to come to my house, we agreed that I'd meet him on the porch of the UP Store. I told Momma I was going to a friend's house and I'd be back at nine. Then I hurried out the back door and down the street and stood shivering outside the store waiting for him to arrive. Off in the distance I saw the faint headlight of his motor scooter coming up the highway.

I ran out to meet him, and I was shocked at how he looked. Though he was dressed warmly, his eyelashes and eyebrows were covered with snow and his face was beet red. I threw my arms around his neck and held his face close to my chest trying to warm it up.

"Come with me. Let's see if the door to the furnace room in the store's basement is unlocked. We have to get you warm," I said. We ran around to the back of the building. When we reached the entrance to the furnace room, I hurriedly ran down the steps to try the door. "It's open! Come on down," I cried out.

Once inside, we stood close to the furnace, rubbing our hands together while we talked. Joe put his arms around me and pulled me close, and I felt his lips against mine. I suddenly felt guilty, like I was doing something that felt good but was wrong. Pulling back, I said, "I've never been kissed by a boy before. Daddy would kill me if he knew I kissed you."

"Why?" he asked. "There isn't anything wrong with kissing."

"You don't know my Dad! If he knew we were doing this, he'd really be mad. I just know he would," I said as tears came to my eyes. "Daddy told me I could never have a boyfriend until I was eighteen years old."

"Eighteen! Is your father crazy or something? No one waits until eighteen to start dating. Besides, how's he gonna know I kissed you unless you tell him?"

"I won't tell him but you don't know my Dad. He has a strange way of finding out things. He says he'd know if I ever disobeyed him just by the look on my face."

After awhile when it was close to nine, I told him we'd better leave. I made him promise to call me as soon as he got home, but told him to hang up if Daddy answered. "I'll know it's you calling and you are at home."

He kissed me goodbye and before he drove off he said, "It was worth coming out tonight just for the kiss."

When I got home, I was shocked to see Daddy sitting at the kitchen table. I went to my room to get ready for bed. Awhile later, I heard the phone ring. Daddy answered, then said, "No one there. Someone must have dialed a wrong number and hung up."

As I drifted off to sleep that night, I found myself thinking back to something Momma once told me when I asked her what "puppy love" meant. I'd overheard this expression when one of her friends was talking about her daughter. Momma laughingly explained, "Puppy love is what they call it when a young girl gets her first crush on a boy. Every young girl experiences this. It's different from when you get older, and you meet a nice man and really fall in love. No one ever forgets their first true love."

I fell asleep remembering Momma's words and wondered if what I was feeling tonight was puppy love, or if it was the real love that I would never forget.

We don't have many photographs of Daddy in his coal mining clothes since he usually changed clothes in the bathhouse before he came home in the evenings. (Coal Camp Reunion photograph)

CHAPTER THIRTEEN
TURN OF EVENTS

EACH YEAR ALL Momma's brothers and their families gathered at our grandparents' home in Rock Springs to celebrate Thanksgiving. Since Momma's sisters both lived out of state, they rarely made it home for the holidays. A couple of days before Thanksgiving, Momma cleaned Grandma's house and starched, stretched, and hung the white lace dining room curtains. As families arrived, the house swelled with excitement, laughter, conversation, and a lot of love.

After everyone was seated before the meal, Grandpa said grace and initiated a toast, drinking wine from crystal wine glasses. Sometimes we kids were allowed a little bit of wine in a glass, so we too could clink the glasses together in salute.

Following dinner, while the women did the dishes, Grandpa sat in his rocker in the living room and gathered his grandchildren around as he read scriptures, carefully explaining each passage as he went along. "Do we have to listen to this, Grandpa?" some of the younger children always asked. He would reply, "What kind of grandchildren do I have who don't want to listen to the words of our Lord? Of course you have to listen." The children then sat back and listened without saying another word.

The adults spent the remainder of the evening visiting while the children sprawled on the floor coloring in the new Christmas coloring books Momma always brought or leafing through the Montgomery Ward and Sears, Roebuck toy catalogues. Christmas carols played softly on the radio. Before we knew it, it was time to pack up, say goodbye and head for home. To us kids, it meant the start of Christmas season.

○ ✕ ○ ✕ ○

On the Saturday morning before Christmas, we bundled up to go to town to search the tree lots for a Christmas tree.

Cars filled the Rock Springs streets, shoppers crowded the stores, and families packed the Christmas tree lots. "Let's go down to the UP Store tree lot, John," Momma said. "Their trees always last longer." The scent of fresh pine filled the crisp December air. Though it was very cold, the sun glistened on the snow.

Carefully we walked through the rows of trees—some set up on wooden stands, others lying on the ground—until we found a beautiful seven-foot blue spruce, Momma's favorite kind. Daddy thumped it on the ground and spun it around to be sure it was nice and full before he took it to the salesman and said, "You can mark this one sold." He threw me the keys to go open the trunk while he paid for the tree. Daddy laid the tree at an angle in the big trunk, pushing it as far back as it would go, though a few feet of it still hung out the back. Then he took a rope and tied the trunk lid to the bumper before he turned to Momma and asked, "Well, Margaret, got any more stops before we head home?"

Before she could answer, I begged, "Daddy, can we please go look around in the Firestone store before we go home?"

"Oh, she's right, John," Momma agreed. "We'd better let the kids see all the new toys down there."

The owners of the Firestone store always stocked many toys for Christmas and seemed to know just what kids wanted. But the main attraction, set up right in the middle of the store, was a huge Union Pacific Lionel train layout, constructed to resemble a little city with trains winding through valleys, along rivers, around houses, past stores, across streets, and over bridges leading into a mountain scene. A two-foot tall glass enclosure encircled the display so children could look but not touch. The display was lit with miniature streetlights, and the trains' engines had headlights and white smoke spewing from their stacks. The owners were like kids themselves and happily answered a multitude of questions from many children. The one question my brothers and I always asked was, "Would you please tell me how much this costs?"

The store also had Erector sets, sleds, trucks, cars, Lincoln Logs, all kinds of board games, a wide assortment of Madame Alexander and Horsman dolls, dolls houses, buggies, Pyrex play dishes, and, this year, the new Red Rider toy rifles.

Before we went home, we went to Union Mercantile for groceries. I stopped at the window to stare at a mannequin wearing a navy taffeta dress with a scalloped neckline. Hand-stitched pearls dotted the bodice. The circular skirt had a sash around the waist, leading to a bow in the back. Black patent leather Mary Jane style shoes were on the floor below it. I pressed my face against the glass to get a closer look. What would it feel like to wear that dress and those shoes?

I ran into the store looking for Momma and when I found her I tugged her arm, "Momma, you have to come see the dress and shoes in the window." She seemed only vaguely interested, so I said. "If I could have that, it's the only thing I would want for Christmas."

When we got home, Daddy and my brothers cut the tree trunk to fit our tree stand. Daddy brought boxes of decorations up from the basement, and he and Momma strung the lights, remembering to put a couple of strings of bubble lights clear around the bottom. Then we all pitched in to hang ornaments; some Momma had carefully packed away each year since their very first Christmas together.

○ × ○ × ○

The week before Christmas, Joe called and asked if I could go with him to the Saturday night movie, if his parents came out to Stansbury to pick me up. Oh, how I longed to go, but how was I ever going to ask Daddy?

I practiced what I would say and how I would answer his questions. I remember the night I finally got up my nerve to ask Daddy like it was yesterday. My whole family was sitting in the living room after dinner when I blurted, "Daddy, I met a really nice boy from Rock Springs at the Saturday matinees. He wants to take me to the holiday movie this Saturday night. His parents said they could drive him out to pick me up at our house and bring me home right after the movie. Please, can I go?" The room went silent while my brothers stared at me wide-eyed, waiting for Daddy to erupt.

With his head buried in the newspaper, and without even looking up, he said, "I've heard you've been spending a lot of time on the phone with this boy. Who is he?" After my lengthy explanation, he said emphatically, "You know how I feel about you having a boyfriend at your age."

Momma interrupted, "Oh, John, what harm could it do if his parents took them to a darn movie?" I held my breath as I waited for his reply, wondering if I'd live through this night.

Sternly, Daddy said, "I'll go along with your mother this time but only since his parents are going along to chaperone. But if I don't like him or his parents, that's it. That'll be the end of it. Do you understand me, Marilyn?"

With Daddy sitting right by the phone, I nervously called Joe. He seemed relieved that I had asked Daddy and survived. We worked out the details.

I got more and more nervous as the time approached. Daddy met the family at the front door and invited them in. Joe's father sat down in an easy chair, while his mother sat on the arm of that same chair. I could feel Daddy looking them over. Then he asked, "Would you like a drink?"

"We don't drink, but we'd have coffee if you have some made," Joe's father replied.

Daddy looked at them in amazement. "What kind of people are you, who don't have a drink during the holidays?"

Everyone in the room seemed to freeze. Momma quickly served coffee, and Daddy had a mixed drink while asking all kinds of questions, which I thought would never end. Finally he stood up and said, "Well, looks like you'd better get going if you are going to get those kids to the movie on time. It was nice meeting you."

I hugged and thanked Daddy, gave Momma a kiss, and could hardly wait to get out of the house and into the car. As we drove off Joe's mother turned to me and said, "You know, Marilyn, we understand your father's concern. After all, you are your father's only daughter, and anyone can see that he's very protective of you. Your parents are nice people."

We sat in the balcony where other young couples sat and midway through the movie, Joe handed me a long, narrow box beautifully wrapped in Christmas paper. Carefully I unwrapped the package under the dim theater lights and took out a pretty red bead necklace. "It's beautiful! Oh, thank you, Joe. This is the first present I ever got from a boy." I felt tears welling in my eyes. Then he kissed me tenderly. Next he placed his arm around me and left it there for the whole movie.

○ × ○ × ○

On Christmas Eve, my brothers and I could hardly contain ourselves. Johnny and Jimmy weren't quite sure about Santa Claus, and I didn't say anything to dissuade them from believing. Long after we'd gone to our bedrooms, I was awakened by a huge thud right outside my bedroom door.

I wrapped my robe around me and opened my bedroom door. There was Daddy, sprawled out on the hallway floor with Christmas packages scattered around him. "Daddy! What happened?" I asked only half awake.

"I fell out of the attic." Then bluntly he said, "Marilyn, there is no Santa Claus. I am Santa Claus! And sooner or later the boys will know this. I had the packages hidden in the attic and slipped on that damn ladder bringing them down."

I laughed and said, "Come on, Daddy. Let's put the presents under the tree before the boys get up. Then we both can get back to bed."

A couple hours later, even before the sun came up, I awoke to the oohs and aahs of my brothers as they opened their gifts. "Get up, Marilyn!" Johnny whispered as he crept into my room. "Santa was here. Come see what he brought!"

I threw on my robe and followed him into the living room, which was lit only by the lights of the Christmas tree. Momma and Daddy came out of their bedroom. Momma went to make coffee while Daddy went downstairs to put more coal into the furnace. "Kids, give me a minute to get some eye opener," Daddy joked as he came up from the basement and took the cup of coffee Momma handed him. Together they sat in the living room with a glow about them, watching us kids open our gifts. Daddy handed Momma a package and gave her a glance I didn't understand as she opened it. She smiled when she saw the beautiful, pink nightgown inside.

I looked around for any package with my name on it; I saw two. I opened one big box and folded back the tissue paper; to my surprise there lay the beautiful navy taffeta dress I had admired. I gently took it out of the box, as if a sudden movement might crumble it to dust, and held it close to my heart. "You just don't know how much I wanted this dress," I whispered.

"Wait till you see what's in here," Momma said as she handed me another package. I tore away the wrappings and took out the patent leather Mary Jane shoes. "Oh, Momma, this is perfect."

"Boy! Look at all this neat stuff we got, Marilyn," Johnny sang out as he and Jimmy fit together sections of the metal track for a new Lionel train. The train set had Union Pacific scrolled on the cars and came with little plastic houses and trees to put on the train layout Daddy had built. They also each got a pair of red leather boxing gloves to use on the punching bag hanging in the basement. The huge new Erector set had enough parts to make a moving ferris wheel. For the family, Santa had left a new Monopoly game and a Chinese Checker board.

"Who's that little box in the tree for?" Jimmy asked. He reached for it and handed it to Momma.

"That's for your sister," Momma replied.

"Me?" I asked. "But I already got the dress and shoes."

Carefully I opened the mystery package which held a bottle of Revlon nail polish, Prince Matchabelli perfume, and a pearl choker that had a red chiffon scarf threaded through the pearls. "Thought that necklace would go nice with your dress," Momma said softly as she watched me tie it around my neck.

"This is beautiful, Momma. I never expected all this."

In the corner near the tree were packages from our grandparents, aunts, and uncles, which contained pajamas, underwear, socks, or money. Each year Aunt Edna made each of us a special Christmas stocking filled with all kinds of little things we could use or eat. Each item was individually wrapped which made five to ten surprise gifts to open.

We ate breakfast and then our friends started coming over to show us what they had gotten for Christmas. Everyone congregated on the living room floor inspecting each other's gifts. Santa Claus always brought

My mother asked me to pose in my new outfit on Christmas Day 1952. I first saw this dress in the window of Union Mercantile and longed for it.

our friends pretty much the same things he brought us. Years later Momma told me it was no coincidence, that the parents got together and coordinated things before Christmas so that we kids received comparable gifts.

○ ✕ ○ ✕ ○

Christmas day, Grandma and Grandpa drove out to Stansbury to have dinner with us. After dinner, many of our neighbors and some of our relatives from Rock Springs came over for homemade eggnog or Tom and Jerrys sprinkled with nutmeg and laced with a little bit of whiskey.

Christmas in the camps was about the only day of the year when our lives didn't revolve around the mine.

○ ✕ ○ ✕ ○

Momma and Daddy belonged to the "Dance Club," a group of Stansbury miners and their wives. This club met throughout the year for dinner and dancing to a local band, usually at the Rock Springs Country Club. The club's New Year's Eve party was a gala event, where men wore suits and women wore their best cocktail dresses.

Members of the Stansbury Dance Club enjoying supper at the home of Slim and Lola Nielson. Left to right: Velma Jenkins, Margaret Nesbit, Velda Ashby, Helen Henderson, Slim Nielson, Bob Henderson, Gene Ashby, Roy Jenkins, John Nesbit.

New Year's Eve was a special time for us kids, too, and it was a long time before any of our parents figured it out. The minute our parents left for the Country Club knowing we were safely asleep in our beds, we hopped out of our beds and met at one of our friend's houses to celebrate New Year's Eve our own way. We brought pop, popcorn, and candy for everyone to share. We turned off all the lights in the house except for the tiny light on the panel of the radio and played "Ghost." We hid, hoping to be the last one found. At the end of the evening, we'd sit around eating our treats and competing to tell the scariest ghost story. Promptly at eleven, we cleaned up our mess and headed for home.

About an hour before midnight, the Dance Club members returned to one of the members' homes in Stansbury to welcome the New Year and enjoy a big meal prepared earlier by the wives.

At the stroke of midnight, the mine whistle blew loudly twelve times, signaling the end of the old and beginning of the new year. The adults hugged and kissed and shouted and then dispersed.

Then the observance of old traditions began. With all the different nationalities in camp, many people had traditions unique to their old cultures. Daddy was a full-blooded Scotsman and the Scots observed the New Year with "First Footing." A First Footer is a friend who first steps over your threshold after the clock has struck midnight. That person was ideally brunette, for it was considered bad luck if the person was blonde. A true Scotsman went from door to door with a piece of coal signifying warmth for the house and a small bottle of liquor symbolizing happiness and merriment. Traditional cake (shortbread) from recipes going into the grey mists of Scottish history was still served in Scottish homes.

The morning after the New Year's Eve celebration, we kids woke our parents early to see what favors they had brought home for us. Daddy's waking words were always the same, "Would you kids please bring me an Alka-Seltzer in a tall glass of ice-cold water?"

○ x ○ x ○

On the night of December 31, 1953, the managers of the UP Store in Winton quietly posted on its front door a notice addressed to "Winton People." The notice stated that after inventory was completed, odd lots of towels, sheets, pillowcases, blankets, clothing, and other merchandise, along with store fixtures, would be sold, on a first-come basis, for next to nothing. This notice shocked everyone. The closing of the Winton mine was fresh in their minds, but never did anyone think the popular general store would close as well. The question on everyone's mind now was "What will be the final outcome for the town of Winton?"

On the first day of 1954, "out with the old and in with the new" took on a real significance for the coal miners who realized that life as they knew it was changing drastically.

○ x ○ x ○

The July 1952 closing of the Winton mine was really the beginning of the end, but we didn't know it or want to know it. The year 1954 would prove to be an even more devastating year for the miners.

On January 31, 1954, Reliance miners were met by a notice hanging on the bathhouse door: the Reliance mine was closed.

The news traveled like a shockwave throughout all the mining communities with everyone wondering what would come next. Miners were scared for their jobs, now more than ever, and angry with the UP for the way in which they notified their employees about mine closures.

Two months later, in March 1954, the Hanna miners found the same "mine closed" notice hanging on their bathhouse door. Miners everywhere were enraged. The local bars were filled with distraught miners vocalizing their disgust with the company they had worked for so hard—for so long—and how it was treating them. The headlines in the local newspaper reflected UP President I.N. Bayless's feelings on what had happened: "No More Mine Shutdowns!" The article went on to say the UP had reached the end of a series of mine shutdowns and that the company would continue to operate the three remaining mines indefinitely: Superior and Stansbury with 320 miners and Rock Springs Number Eight Mine with twenty-four miners.

○ × ○ × ○

While the year began fraught with sadness, depression, and uncertainty, the coal camp people went on with the same fortitude they had exhibited for many years. Whether they chose to ignore the inevitable or to savor everything they had once enjoyed, camp life went on as if it would continue forever.

On the heels of the Reliance and Hanna closures, the miners confronted yet another hardship. In April 1954, the operators of the Lincoln-Alliance bus line at Rock Springs discontinued transporting miners between Rock Springs and the coal mine at Stansbury. Operators Florian Anselmi and Erneste Segna had run three roundtrips daily carrying miners to and from their jobs. After the shutdown at Reliance, the service dwindled to two roundtrips a day to accommodate one shift, but now not even one shift worked consistently. Thus, they could no longer afford to run their buses.

Some felt the business community of Sweetwater County had been apathetic to the mining industry and miners. However, once businesses realized the ramifications of closing mines, they began feverishly to seek something else to supplant the income they had derived for years from the mines and mine families. They were now faced with the same fears as the miners: loss of their livelihood.

At the end of the school year in 1954, only fifty-two students were enrolled in the Stansbury Elementary School, and only three of the original six teachers remained: Zoe Lee, Judy Pryde, and Dovey Hanley.

A notice went out that Stansbury Elementary School would be closed before the school year began again in September 1955. When the people of Stansbury heard this news, they realized that they had better make plans to leave Stansbury, because Stansbury was leaving them. A feeling of gloom hung heavy over every family in camp.

○ × ○ × ○

Tippy was waiting for me when I got off the school bus from Reliance on the last day of the school year. Feeling melancholy, I slowly walked over to the mine office and sat on the steps to wait for Daddy so we could walk home together. As I waited, the breeze blew against my face, and I stared off in the distance watching a train make its way along the railroad tracks. The sun was so bright, I had to hold my hands over my eyes like a visor. It was then I noticed the shiny new engine pulling the train coming into camp. Something was different about it; it didn't have black smoke billowing out of its stack. I stared in amazement, trying to figure out what was going on.

The door behind me suddenly opened and Daddy walked out, "Well, look who we have here," he said as I stood up to greet him.

"Daddy, look down there at that new engine coming into camp. I've never seen one like that before. What's different about it?" I asked. As we stood hand in hand looking off in the distance, a strange look came over his face like a veil and he solemnly replied, "That train is being pulled by one of those new diesel engines that's going to put your dad out of a job."

"What?" I exclaimed looking at him in alarm. "How could a train engine put you out of a job?"

"That engine doesn't need coal. It's diesel-powered." We stood quietly watching in awe without saying a word to each other. I'd never seen this look on his face before. It scared me. He broke the silence. "We better get going. Your mother is probably wondering where the heck we are."

Daddy never shared his fears about the mine with us kids, nor did Momma. At the dinner table that night, he didn't say one word about what we'd seen. Our conversation was centered around the proposed closing of the elementary school and how bad we felt about it. But at the same time, Johnny was excited about going to the bigger school in Reliance and riding the bus.

I looked across the table at Daddy and wondered what he would do if that diesel engine did put him out of a job. Where we would go? I loved him so much, was so proud of him, and never ever worried about him when the mine whistle blew because he was a skilled miner who knew the mine so well and knew how to do his job safely.

CHAPTER FOURTEEN

SPRING AT LAST

PRACTICALLY EVERY boy and girl in the mine camps was a member of the Boy or Girl Scouts. Kids looked forward to the day when they bought their first scout uniform at the C.A. West Store.

The part of Scout program which drew the most enthusiasm was the UP first aid program for kids twelve and older. Through this program, co-sponsored by the US Bureau of Mines, young people were taught the same first aid procedures as the miners.

The sky hung heavy over us that winter and the closures were never far from our minds. Except for school, all I could think to look forward to was the annual first aid contests in June. Competition would be keen, judging exact and efficient, and only a few points would make the difference between first and last place at the contest.

The Rock Springs first aid contests were held in conjunction with the Annual Meeting of the Old Timers' Association as part of their anniversary celebration. Every year on June 15, people from the surrounding mining camps came to either participate or watch this day's activities. My love of the event began when I was ten years old and drew the name of the lucky miner who won a brand new car at the meeting. It was at this time I first met I. N. Bayless, who was then vice-president of Union Pacific Coal in Wyoming.

Mr. Bayless oversaw all mining operations in the Rock Springs area. Momma knew him from when she worked at the UP Store when she was eighteen years old. Daddy knew him from his visits to the mine. He was referred to by all the miners as "the UP big shot." But to me, he was just a nice man who came out to Stansbury, always dressed in a

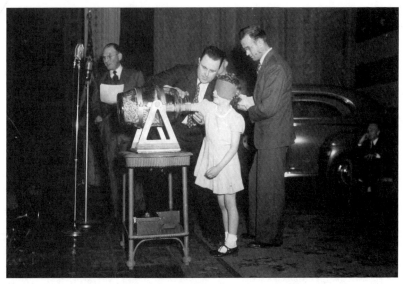

I drew the name of a miner who won the car shown in the background during an Old Timers' Meeting in Rock Springs.

three-piece suit, who put his arms around me and my brothers whenever we saw him at the Stansbury Mine office.

○ ✕ ○ ✕ ○

My interest in first aid was spurred by the story Daddy told us about the time in 1940 when his teenaged niece, Jane Wilson, along with other students, was unexpectedly called upon to administer the first aid they had learned through this program. A bus transporting thirty miners from Winton to Rock Springs skidded on the slippery road and turned over into a barrow pit just south of Winton. Several of the men were suffering from shock, bleeding, bruises, and other injuries.

Along came the high school bus carrying many of the students who were members of the Winton Girl Scout First Aid Team. They assessed the situation and administered first aid until ambulances arrived. Their assistance endeared them to these injured men. The following summer this same team of young ladies won first place at the annual first aid contest. Each time I listened to Daddy tell the story, I longed for the day that I would be trained to render assistance in time of an emergency.

○ ✕ ○ ✕ ○

Once a month a group of miners got together on a Friday at a bar in Rock Spring to play poker. A couple days before Daddy went, he'd ask us kids if we wanted to get in on the "double or nothing" game. We immediately went to work after school each day selling empty pop bottles and doing odd jobs to make the money to give to him to bet.

At supper on Friday night, we proudly handed him the money we had earned and eagerly awaited the results. The next morning, we ran to the dining room table where there was a pile for each kid. If he'd won, our money was doubled. If he'd lost, there was a sack of Planters Peanuts and a Hershey candy bar for each of us.

o x o x o

One Monday after school, the kids in our neighborhood got up a game of baseball in our backyard. Around four o'clock, we heard the shrill sound of the mine whistle blowing. Everyone knew that many of the day-shift workers were still down in the mine. We all ran as fast as we could toward the mine portal, dodging mothers backing their cars out of their garages to go to the mine as well.

When we got to the portal, everyone stopped and stood in silence on the hill above the mine waiting to hear what had happened and who was hurt. The silence was deafening as the mantrip came out of the mine.

"Who got it today?" a woman hollered out in a trembling voice. All eyes were focused on the mine entrance and the commotion around it. At that moment, I prayed silently that nothing had happened to Daddy. The names of the injured miners quickly moved through the crowd. I breathed a sigh of relief when I didn't hear Daddy's name.

I spotted him helping carry a stretcher to the ambulance. Miners who had assisted in the rescue were covered in blood. By listening surreptitiously to the adult conversations, we learned there had been a minor cave-in and that a miner from Rock Springs had sustained a broken back, while two others had minor injuries. I thought how horrific it would be to get news that my father had been injured in a mine accident, or worse yet, had been killed. The mere thought of anything happening to Daddy brought fear to my heart.

We had a late supper, and Daddy explained the man with the broken back had come over from the old country and was working as many hours as he could to save enough money to bring his family to America. Daddy said he'd be going to the hospital to check on the injured man just as soon as we finished eating because, as he put it, "the poor guy doesn't have anybody here." After he got home from the hospital that night, he sat down and wrote a long letter to the man's family, assuring them that the injury was not fatal, and that their loved one would soon be able to return to work.

In the days that followed, Daddy seemed to grow more short-tempered over the smallest things. He also became relentless in asking me where I going and with whom.

○ × ○ × ○

Around this time, he and Momma decided it was time to talk to me and Jimmy about the facts of life. Daddy, carrying a dark blue book in his hand, took Jimmy into the kitchen and closed the door, while Momma, carrying a similar book, took me into my bedroom. My brother and I looked at each other wondering what the heck was in the blue books that was so ominous.

I sat next to Momma on edge of my bed as she read through the pages, which also contained ghastly pictures of diseases I had never heard of before. My eyes felt as big as golf balls when she finished and asked if I had any questions. The first thing that came to mind was, "Do teachers do that kind of stuff?" Her reply shocked me.

Before going to bed that night, I turned my phonograph on low so I could listen to 45 RPM records while I fixed my hair. Then I set curls in my hair with bobbypins. Suddenly, Daddy burst into my room and angrily announced that from now on I would have to have my light off and be in bed by ten. I looked up at him in shock. Then he angrily turned off my phonograph and switched off my bedroom light. The container of bobbypins spilled across my bed. I fell asleep crying while lying among the scattered pins. The next morning I told Momma what he had done. She brushed it off with, "Oh, your father has a lot on his mind these days. Just do as he says to keep peace in the house."

○ × ○ × ○

This year for the first time, my parents allowed me to go on the bus with my friends to the district basketball tournament in Green River. A school bus transported students and chaperones who wanted to attend the Friday and Saturday sessions. We had great fun watching the games and when our school team wasn't playing, we walked over to the island along the Green River to roller skate in the National Guard Armory. When the last game was over each evening, we promptly met back at the bus for our trip home.

The minute I got home, Daddy started quizzing me, asking if I had been with any boys. In particular he wanted to know if the boy he let me go to the movie with was there. "Yes, Daddy, he was there, but I wasn't with him," I replied. Surprisingly, he seemed to believe me.

As soon as I could, I called Joe and told him I couldn't see him until I got older because my father seemed too upset by it. At first he couldn't understand but we talked until he understood my predicament. A few weeks later at a Saturday matinee I saw him sitting with his arm around a girl I didn't recognize. He sheepishly smiled at me as I walked by. It hadn't taken long for him to replace me.

○ ✕ ○ ✕ ○

The school district began hauling things out of the Stansbury school building to transport to Reliance Elementary. Camp kids sat on the playground equipment watching one truck after another being loaded with things from the building: student and teacher desks, chalkboards, bulletin boards, and the piano. Sometimes, when the movers let us, we helped carry boxes of books and school supplies to the trucks. Then plumbers removed the drinking fountains and bathroom sinks, stools, and fixtures. As each day passed, the building looked more desolate. Then the workers boarded all the windows and doors of the building. Our schoolhouse now sat barren.

At the dinner table each evening, we talked about what we had witnessed. Momma and Daddy sat silently and listened, never interjecting any comments. Then one night Daddy said the boarding house was going to close. The family who leased the building had announced they would be leaving Stansbury right after their son and daughter, who were both seniors, graduated from Reliance High School. When Momma heard this news, she cried. We each mourned the changes to our community, never wanting to admit that they were inevitable.

Immediately after supper that night, my brothers and I ran down to the boarding house. The owners, Ann and her husband, Jim, were sitting on the front porch visiting with disheartened boarders who would now have to find another place to live. We listened as they reminisced. Jim told how hard it had been holding down two full-time jobs, working every day in the mine and coming home to help Ann with all the heavy work. He helped pack all the lunch buckets, hauled in the wood and coal for the furnace and hot water stoves, cleaned all the boarders' rooms with the help of a cleaning lady from Rock Springs, and mopped all the floors. The work was so time-consuming that five years previous, Jim had started working solely at the boarding house.

Jim and Ann's daughter Jo Ann did the dishes and set the tables for the evening meal, but her father never let her go into the men's rooms to help clean or gather bedding to be laundered. He felt that was no place for a young girl.

After school, their son, Billy, peeled all the potatoes and onions for the evening meal and then washed all the heavy pots and pans, of which there were many.

The boarders also fondly remembered times Jim and Ann joined them in the boarding house lobby to play cards. Ann loved pinochle and played as often as her busy schedule allowed.

When my brothers and I got home that night, we lamented to Momma and Daddy about how much we'd miss this friendly, outgoing family. They always had an extra place at their table, especially if you were hungry for a slice of Ann's famous dessert, *potica*.

○ ✕ ○ ✕ ○

Later, when Daddy popped his head into my room to say goodnight, I asked, "Daddy? What's going to happen to us? Are we going to have to move, too?"

He paused for a moment and then said, "Probably, but not for a while yet. But don't worry your pretty little head off. Your mother and I will always be here to take care of you kids no matter where I work or where we live."

CHAPTER FIFTEEN
FIRST AID CONTEST

THINGS FELL like dominoes once the Reliance Mine closed: the Hanna mine closed, the Stansbury school closed, the bus line quit running, the boarding house closed. Uncertainty was the only thing certain.

The miners at Stansbury and Superior, where the mines still operated, could see the end was near. Kids looked forward to summer, but the summer recreation program—including the baseball league—had closed down. The only part left was the Friday swim at the Rock Springs High School swimming pool.

One hot summer morning after breakfast, I went over to Carolyn's house to see if she wanted to come over to run through the sprinklers in my yard. Carolyn met me at the back door with tears streaming down her cheeks and flung her arms around me.

"Oh, Marilyn, Daddy says we have to move," she sobbed.

"Why now? The mine is still open," I replied in shocked disbelief.

"Daddy thinks he'll soon be out of a job if we stay here. I heard Momma crying all night. None of us want to leave."

Tears swelled in my eyes, too. "I've got to talk to your parents," I uttered as I hurried past her, somehow expecting that I could change their minds. Her mother and father sat deep in thought at the kitchen table. "Is it true? Are you guys really moving?" I blurted out.

"Got to do something," her dad replied. "Not going to be anybody left at Stansbury besides ghosts." He said he'd been calling around looking for work in places like Colorado and California where they had family. As soon as he found any kind of job, they'd pack up and go.

"But, you won't leave before the first aid contest, will you?" I asked.

"Marilyn, I hate like hell having to leave as much as Norma and the kids do. But I have a family to feed. I don't know what we'll be doing tomorrow let alone the day after. I'm just taking it one day at a time right now."

"But if you move before the contest, would you let Carolyn stay with my family at least until after the contest?" I begged. "She's part of our team! We need her."

"We'll see," was all he said.

I felt all alone as I ran back home. The camp seemed bleak everywhere I looked. A strange feeling came over me, one I could not describe—and unbeknownst to me it would linger with me for many years. I felt like a part of me was being carved away and try as I might, I couldn't get it back.

I burst through our back door and ran to Momma who was standing by the kitchen counter making sandwiches. "Momma, I've got really bad news. Carolyn is moving away. Oh, Momma, what am I ever going to do without her?" I sobbed as I wrapped my arms around her waist and laid my head on her shoulder.

Momma said she and Daddy had been trying to keep us kids from worrying. It shocked me when she admitted Daddy was applying for other jobs and before long we'd be moving too. She said she didn't see any other end to this. Then she tried to find the positive and confessed she worried about Daddy every day he went to work, so in some ways closing the Stansbury mine would be a blessing.

○ × ○ × ○

The day of the first aid contest was drawing near. Fortunately, our team was intact except for the girl who played the part of the accident victim. She'd moved but had been replaced. Two weeks before the competition, Daddy announced we'd all go to the UP store in Rock Springs to order our team uniforms. We chose white sailor hats with our first names embroidered across the front, white short-sleeved shirts with red ribbing around the necks and sleeves, silver concha belts, and jeans. We wore our own shoes since the company did not pay for shoes.

The day of the competition began with a parade of the teams, which marched along North Front and K Streets, ending up at the UP

The first aid teams marched as units in the Old Timers' Day Parade in the 1950s.

Store. Floats depicting the mining industry and other organizations, cars carrying mining dignitaries, and children riding bikes decorated with crepe paper joined in the parade.

After the parade, the UP sponsored a sit-down luncheon at the Masonic Lodge for all first aid contest participants and spouses. UP hired local women to prepare fried chicken, mashed potatoes, vegetables, salads and a variety of desserts. Then we went to the Old Timers' building. Inside the hall thirty-two flags hung, representing the nations of the 693 Old Timers' members from Rock Springs, Reliance, Stansbury, Winton, Superior and Hanna. The hardwood floor was polished to a high luster. Our team waited along the outer wall with the others, and I began to wish I hadn't eaten so fast as my stomach was feeling funny.

The hall was filled to capacity. After an introduction of dignitaries and an explanation of the rules, a bell rang signaling the beginning of the contests. First to compete were the young women's teams, followed by the boys, and then the men. The teams stood in a large circle in the middle of the room waiting for judges to hand the injury problem to the captains. Each captain read the problem aloud while the team members listened carefully so that simple errors weren't made—like giving artificial respiration to a patient with broken ribs. Each team had a different problem to avoid any cheating by copying other teams' actions.

The Old Timers' Day luncheon was a huge affair. Johnny Nesbit is shown in front, third from left.

I desperately wanted my team to do well in this competition to please Daddy because he had trained us, but also to show that even though we were young we could be of value in an emergency. During a mine accident the whole community needed to pull together, and I wanted to show we were good enough to be a part of that.

Our division took the floor, when the judge walked over and handed me the problem, my hand visibly shook as I took the sheet. I quickly scanned it, then walked to my team and read it out loud. Together we decided how to proceed.

Our patient, a man, had fallen off a high rock ledge outside of town and was lying face down and lifeless on the ground below. He was conscious and complaining that he could not feel his legs. He was bleeding profusely from a cut on his right calf.

Though it has been over fifty years, I still remember how we proceeded. We decided our patient likely had a broken back, but our first concern was to stop the bleeding. I applied a tourniquet tightly around the pressure point above the wound, and we checked the tourniquet every ten minutes for proper blood flow.

Then we laid the back splint from our locker alongside the patient. Without turning the patient over, we carefully raised him by putting

The team moves the "patient" onto the stretcher during the first aid competition.

our arms under him from each side. We slipped the splint under him and secured him to the splint with cotton cravat bandages.

When the whistle sounded signaling the time had expired, we stood by our patient as the judge critiqued our work. He complimented us on the use of the tourniquet but said a few of our square knots were not tied correctly and slid when he pulled on them. This infraction cost us points.

They announced the winners in reverse order at the end of all the competitions, People screamed and applauded as each place was announced. In our division, our team was the second prize winner, and the other teams erupted in a loud applause. I said to Daddy, "See how many people are cheering for us!"

He grinned at me and explained, "They are just cheering because now they know they have a better chance for first place, since your team took second."

I felt like a fool.

But if we couldn't get first we were happy to get second because the prize was great. Each member received a Kodak camera, which I immediately started using to take pictures of the event. Daddy's team took first place and each team member won $40.

○ X ○ X ○

I took a photo of the five members of my first aid team with my new camera. Standing: Sherry Jenkins, unknown chaperone, Gloria Fabiny, chaperone. Kneeling: Unknown, Darlene Fabiny, Carolyn Pecolar, and Carrie Palcher.

After the competition, my parents dropped my brothers off with my grandparents, me off at the New Grand Café, and then they joined friends for dinner at a restaurant followed by dancing at Giovolies' Bar. Daddy gave me money to go with my friends to get a hot breaded veal sandwich, and afterwards, all us kids went to the Eagles Hall—the first time I'd been allowed to go to a dance without my parents. When Daddy dropped me off at the café, he gave me strict instructions not to leave the dance hall until he arrived to pick me up.

The hall was packed with kids. The band was playing "Secret Love" when I walked in. I was afraid to be asked to dance, but also afraid that I wouldn't be and I'd be left standing on the sidelines. Then a boy I'd never noticed before asked me to dance. Within a few minutes I found myself able to follow his lead. We talked as we danced, and I learned he was Carl Tomassini and that his father worked with my father. I danced with different partners practically every dance, but it was Carl who taught me how to jitterbug, and once I got the drift of it, I could have done the jitterbug all night.

Near the end of the evening, I noticed Daddy walking slowly across the dance floor toward me. My friends looked at each other puzzled at why he'd come in. "I thought you were gonna meet me outside after the dance was over," I said.

"Well, I thought I'd have the last dance with my daughter," he replied as he took my hand and led me onto the dance floor. I have to say he was the best of all the dancers that night. When the music stopped, we walked off the dance floor and out of the building through a throng of noisy teenagers.

○ × ○ × ○

The next day, my head and my heart both dropped when I saw the big truck parked in Carolyn's driveway. I hurried over to help pack and as the house became more empty our voices echoed in the empty rooms.

"I'll never, ever forget you," I told Carolyn as she and her sisters got into their family car to drive behind the moving truck their father was going to drive. "I'll write to you every single day, so soon as you get my letters, write right back, okay."

"Cross my heart I will," Carolyn promised.

The wind swirled through the camp, kicking up sand as their vehicles pulled out of their driveway. I jumped on my Schwinn bike and followed behind them as they slowly drove away along the winding dirt streets leading out of camp. When they reached the paved highway, I just couldn't go back home, so I followed behind their vehicles the mile down the highway to the Winton cutoff. I was crying so hard I could barely see the highway or their vehicles ahead of me. The depth of my sadness was indescribable. A part of me left with them that day.

Sometimes through my tears I caught a glimpse of Carolyn and her sisters, who were crying as hard as I was, as they leaned over the back seat of their car, peering out the back window waving to me. At the Winton turnoff their mother stopped the car. For a moment I hoped against all hope that she'd changed her mind and they were staying.

I pulled my bike alongside their car and leaned inside the back window. Carolyn's face was red and swollen from crying. I wiped my face and nose on my sleeve and saw even their mother was crying with her head bent over the steering wheel.

I didn't need to ask. Nothing had changed; they were going. I gave Carolyn one last hug. After a moment, her mother lifted her head and said, "Come on, girls. We can't drag this out any longer." I stepped back from the car and sat on my bicycle seat balancing with one foot on the ground and watched their vehicles until they were out of sight and all that was left was the sun glistening off the barren prairie.

With cars passing me on the well-traveled highway, I rode home as a kaleidoscope of memories clouded my mind. Carolyn was like the sister I never had. We shared everything: M&Ms, popsicles, childhood diseases, dolls, music, and our treasured movie magazines. Everywhere we went, either her parents or mine took us, we wore the same style clothes and shoes, shared confidences, diary secrets, hopes, and dreams. Her leaving was a drastic and inexorable change. I wondered how my life would be now that she was gone. As I write this, I can still feel the sadness of that day.

My brothers were in the front yard when I got home. "What happened to your face?" Johnny asked. "There's muddy stuff all over it."

"I followed Carolyn and her family down to the highway. I guess the sand was blowing so hard it made my eyes water. I'll go wash it off," I replied.

As I turned to walk into the house, he blurted, "Guess what, Marilyn? Something terrible! Now the Hendersons are moving away. Pretty soon we're gonna be the only family left in Stansbury."

I shoved my bike up against the fence and ran into the house. "How much more can I take?" I shouted.

Not even stopping to wash the muck off my face, I went straight to my bedroom, closed the door, and threw myself across my bed and sobbed until I fell asleep.

That night at supper Momma and Daddy did all they could to console us about our friends moving. I don't remember exactly what they said, but somehow that conversation gave us the strength to accept the fact that one by one our childhood friends would be leaving Stansbury and leaving us.

Back: Joe Mecca, Johnny Nesbit, Deforest (Slim) Nielson. Kneeling: unidentified, Eugene Ashby, Carlos Tarufelli. The men made up one of Daddy's first aid teams that received first place.

○ ✕ ○ ✕ ○

A couple weekends later, our family went to Green River to attend the big United Mine Workers' Union picnic, which was held each summer on the island in the Green River. People parked cars everywhere, surrounding long rows of picnic tables covered with colorful oilcloth tablecloths. Young and old alike played horseshoes and baseball, and some even swam in the Green River. Others went rollerskating in the National Guard Armory or swimming in the city pool both nearby.

When we got home that night, Daddy surprised my brothers and me when he said he was going to let us spend the rest of the summer in Salt Lake with Aunt Ann and Uncle Zeke. I guess he wanted to get us out of Stansbury so we didn't witness first hand our friends moving away and he didn't have to endure us moping around. This time, even Johnny wanted to go to Salt Lake.

"But just remember, they may treat you like you're their kids, but you're not their kids," Daddy snapped, "and don't you ever forget that." Then he softened, "It won't be the same with all of you gone at the same time. I like my kids to be home when I get off work each day."

This time we got to take the train to Salt Lake and soon we were settled in with Aunt Ann and Uncle Zeke. Aunt Ann signed us up at the YMCA and YWCA for swimming, tennis, and crafts. She even signed me up for water ballet and ballroom dancing classes. She taught us to read the city bus schedule and to recognize the landmarks so we could ride the bus downtown to the recreation center by ourselves.

One Saturday afternoon, Aunt Ann and Uncle Zeke took the boys downtown to shop for their school clothes. Later that evening their friend who was a women's clothing salesman brought over boxes of sample items that were my size. I was shocked beyond words when Uncle Zeke bought me every single outfit: dresses, coats, suits with matching blouses, and twelve swimming suits. As a coal miner's daughter who was used to getting five new outfits a year, it was hard for me to even imagine what a person did with all those clothes. I remember that night to this very day.

Throughout our stay, Aunt Ann taught me many things including to knit, crochet and sew on her Singer Featherweight sewing machine. After she explained all the workings of the machine, how to maintain it, and how to select the correct size needle for different types of fabric, she taught me how to use a pattern and sew my own clothes. One day while I was sewing she came to me and said, "When you go home, I'm going to let you take the Featherweight back with you."

"Whoa!" I jumped up and threw my arms around her. " Wait until Momma hears this!"

Soon the summer was over. My brothers and I loved traveling on the train, walking through the train cars, and eating in the dining car after the conductor came through the cars calling out, "First call. First call for dinner."

When we got off the train in Rock Springs, for a minute Momma and Daddy didn't even recognize us, because we were dressed so nicely, and I had to call out as I waved one arm above my head, "Daddy, we're over here!" They looked like they were as glad to see us as we were to see them. My brothers and I certainly looked different; we looked like city kids. Jimmy and Johnny wore nice dress slacks, short-sleeved dress shirts, and brand-new shoes. I wore a grey suit with matching hat and heels. It was the first time I had ever worn a hat anywhere other than in Salt Lake. I felt like a model.

"I should put you kids right back on the next train," Momma said as we walked to our car. She told us there was a polio epidemic in Rock Springs far worse than what we'd had last summer when a young Rock Springs girl first came down with the crippling disease. Then Dr. Crocker, an optometrist who was only thirty-seven years old, died when he came down with the disease. This dreadful illness either paralyzed or killed. Many people didn't survive the iron lung longer than three days. Now people were speculating that polio was spread in swimming pools, and for that reason Momma didn't want us near a swimming pool.

I remember how Momma worried when one of us kids got the mumps, measles, or chicken pox but that was nothing compared to the way she worried about polio. Everyone in the camps was shown films at the community hall on polio and the iron lung machine.

The outbreak was so bad that Dr. Muir moved into the Park Hotel in Rock Springs for a while rather than take the chance of bringing the disease home to his wife and young children. He spent night and day at the hospital.

One night when I was at the hospital sitting in the waiting room waiting while my mother visited a sick friend, I saw Dr. Muir coming down the hall headed for the room that housed the iron lungs. I went up to him and asked him to let me see what an iron lung looked like. He took me down the hall to a room where he gave me a gown and mask. Then he walked me to the door of the room where the iron lungs were. Quietly we stood looking through the outer windows of the doors while he explained just how the machines were helping the patients breathe. I couldn't see any faces of patients inside but I imagined them encased by the machine and peering up through the glass window on top of each iron lung. All I could hear was the rhythmic hissing of air being compressed into the machines. I was scared to death of polio.

○ × ○ × ○

After having spent most of the summer in Salt Lake, it was hard to get used to being back in Stansbury again. Even the air smelled different. Stansbury air smelled like mine water. As soon as I woke up, I could hear the wind blowing outside and my heart sank. The one thing I hated about Stansbury was the wind, which never seemed to blow hard in Salt Lake. Slowly I got out of my bed and walked to my bedroom windows to pull up the shades. Outside the wind was swirling sand in a whirlwind motion through the camp making everything look bleak and dreary. I glanced down at the windowsill, which was layered with sand. At that moment, I hated being back home and wished I lived somewhere else.

When I sat down at the kitchen table, I looked over at Jimmy. As our eyes met, I knew he was thinking about the same thing—the stark difference between Stansbury and Salt Lake.

As Momma served our breakfast, she told us how much she missed the sound of our laughs while we were away. Then Johnny started spouting, "I really missed you, Momma. Aunt Ann made us take a bath every single night, and even made us scrub our feet and elbows with a brush. When their friends came over, they called us "miner kids.""

"Oh, Johnny," Momma replied laughingly. "Was that so bad, Son?" He just hung his head and didn't say any more.

After the dishes were done, Momma drove us into Rock Springs and went straight to Safeway. We followed her down each aisle as she checked items off her grocery list. When we got up to the checkout counter, she did a quick mental calculation of the cost as she placed the groceries on the counter. When the clerk gave her the total, she said, "Oh, I guess I figured wrong. I don't have enough money. Let me see which items I can put back."

She set some items aside while those standing in line behind us gave us annoyed looks, which didn't seem to bother Momma in the least. I wanted to die I was so embarrassed, and I could tell the boys were, too.

○ × ○ × ○

After supper that night, my friend Geraldine stopped by. I showed her my new clothes and told her about Salt Lake; then we decided to ride bikes. At the schoolgrounds, we watched some boys who were riding horses. Soon they rode over and asked if we wanted to go for a ride.

One boy reached down and pulled Geraldine up to sit on the saddle behind him and the other gave me his hand as I put one foot in the stirrup and climbed up behind him. I had so much fun sitting on the horse as it trotted around camp with my arms wrapped tightly around the boy's waist.

That evening after supper, while my family was sitting in the front yard, I told Momma about Don, the boy who had taken me horse riding. "My lands, what a coincidence," she exclaimed. "When you were born, his mother shared my room in the maternity ward. He was born the same day as you. Isn't it funny how things work out? I remember saying to her wouldn't it be something if our kids grew up and fell in love with each other." Now curious she asked, "What kind of boy is he, Marilyn?"

"Oh, Momma. He is the nicest boy I have ever met and he's so handsome. He's tall with dark, wavy hair, and pretty blue eyes. We'll both be in the eighth grade when school starts. He lives in Reliance."

Don and his friend came to Stansbury practically every weekend for the rest of the summer. They'd either ride their horses or bikes, or just walk the mile shortcut between Reliance and Stansbury. Many times Geraldine and I walked them halfway home and stopped at Red Rocks to explore the rock formations.

On Saturdays, I started meeting Don at the theater matinee and later in the evening at the teenage dances held at the Eagles. I swore my brothers to secrecy so that Daddy wouldn't find out. Momma was glad I'd met Don, but I wasn't sure if Daddy would feel the same way.

When school started the only time we were apart was when he went home for lunch. We both loved school and sometimes even competed against each other to see who could get the best grades on exams. We talked about college and our desire to earn honor scholarships when we graduated.

I realized how dependant I'd become on Don when he came down with rheumatic fever and had to stay home from school for several weeks. I ran to his house every weekday at noon to see him after I'd eaten my lunch.

o x o x o

One Saturday afternoon, a carload of kids from Reliance drove up in front of my house and honked the horn. Daddy went outside to see what they wanted, and to tell them if they wanted to talk to me to come to the door, never sit in a car and honk. When he came back into the house, he told me I could talk with them in the yard, but not to get in the car with them. I ran outside to see who they were.

One by one they piled out of the car, including Don. I was relieved that he was finally well enough to go out.

We all sat on the lawn in my yard talking and laughing. I didn't show any outward affection toward Don for fear Daddy would come running out of the house and ask everyone to leave. He had done that on several occasions when someone he didn't like came to see me, even if it was a girl. I hated it when he did that.

Pretty soon Daddy came out of the house and asked if anyone wanted to play a game of croquet. Momma brought out lemonade and homemade cupcakes for refreshments while we played.

Later Momma made hamburgers and French fries, and we all sat around the yard telling stories.

Everyone had a story to tell that night and before we knew it, it was dark. Don and his friends thanked my parents for such a nice time. I walked them back to the car and Don said, "I don't know where you get the idea your father is so strict. Your parents are really great."

"Yes," I said. "Daddy must have liked you."

CHAPTER SIXTEEN
A CHILL FALLS

NINETEEN FIFTY-FOUR was a devastating year for the coal industry in Wyoming. A vast amount of coal lay underground, but there just wasn't a big market for it anymore. Rumors began spreading throughout the camps that Union Pacific had 190 diesel engines on order.

Sometimes after we'd all gone to bed, I could hear Daddy talking to Momma about moving. My ears really perked up the night I heard him mention West Virginia. The next day I told him I'd overheard what he'd told Momma, and I wanted to know what West Virginia was like. He said West Virginia was beautiful with green trees as far as the eye could see. When he reminded me that the movie I liked so much—*Raintree County*, starring Elizabeth Taylor—had been filmed in West Virginia, I was impressed. "But don't be telling your friends we're moving because right now I don't know where we'll be going. Everything is still up in the air, but coal mines are still hiring back there to fuel the steel mills."

○ × ○ × ○

When they officially closed the boarding house, the company boarded up the windows and doors and hung big NO TRESPASSING signs on each side of the building.

Each morning, when my brothers and I walked to catch the school bus, we walked by the empty school building which was starting to show signs of disrepair. Then, one morning Johnny shrieked, "Look, Marilyn! Someone stole some of the playground equipment." Everyone surmised it happened in the dark of night. When we told Momma what had happened, she warned my brothers and me if we wanted to keep our bikes, we'd better be sure to put them in the garage each night.

○ ✕ ○ ✕ ○

Riding bikes around camp wasn't as much fun anymore since there was so much gloom everywhere we looked. The camp kids desperately needed something to do and somehow we came up with the idea of turning the abandoned boarding house into a secret hideaway. Under the cover of darkness, with flashlights, we ignored the no trespassing signs and broke into the boarded up building through a basement window in the back. Once inside, we began staking out the rooms for individual cabins. Night after night, we hauled blankets, orange crates, candles, dart boards, and our roller skates through the basement window. The dining room made a perfect place to roller skate. After supper each evening, the inside of the building came alive with camp kids sneaking in and out. We kept our voices low so we wouldn't be noticed by passersby. We swore each other to secrecy so that our parents didn't find out anything.

Early one Saturday afternoon we all met at the boarding house, crawled in through window and lit up the darkened dining room area using candles and flashlights and proceeded to have a great time roller skating. Suddenly we heard a commotion outside and ran to peer out through the openings in the boarded up windows. No one uttered a word as we looked down in amazement at all the men and company trucks that completely surrounded the building. Soon a man angrily yelled through a bullhorn over and over, "Everyone inside, exit this building immediately!"

We were so scared our feet felt like they were stuck to the floor until one of the kids said, "We'd better get out of here before they start shooting." Then we hurried to our only exit—the window.

One by one we all scrambled out the basement window, and the bullhorn ordered us to line up against the building. Practically every kid in Stansbury was in that lineup. We looked like condemned men in a firing line. I froze when I saw Daddy standing next to the man with the bullhorn, and I knew my brothers and I were in a whole lot of trouble. I had never seen Daddy look so angry as he glared straight at me and said, "What the hell were you kids thinking? This is company property. You can go to jail for trespassing."

As soon as we got home, our punishment was swift and effective. Momma told us Daddy was so mad that he didn't spank us because he'd probably kill us if he did. Instead, all of our privileges were taken away for two weeks: no allowance, no friends could come over or call, no Saturday matinees, and we were confined to the house after school and on the weekends. As the days went by, Daddy hollered at us even more when he learned just how much stuff we had packed into the boarding house.

o x o x o

After Thanksgiving, Daddy and Momma seemed to have moved past the boarding house fiasco, but many things were different even during the holidays. Johnny's Christmas program wasn't held at Stansbury as it had been for years but rather at the Reliance High School auditorium. The union members still made sure Santa Claus came to the program and handed out Christmas stockings with candy, oranges, and apples. But, for the first time, the miners did not receive a Christmas bonus.

The dance club my parents belonged to for so many years didn't have a New Year's party at the Country Club this year because so many members had moved away. Momma and Daddy decided to have a few friends come over to our house for a New Year's get-together. A friend's parents took her and me to Rock Springs to attend a New Year's dance at the Eagles where I had met Don.

When we came home our house was filled with friends, eating and drinking, all having a good time. Daddy had rolled back the wool living room rug so couples had plenty of room to dance to the music playing on our Motorola. We kids got to drink as much pop as we wanted while laughing and listening to Daddy and his friends, arms around one another's shoulders, harmonizing to some of their favorite songs.

At midnight the mine whistle blew, and everyone welcomed in the New Year while wondering what it would bring.

o x o x o

Daddy started coming home from work later and later due to the long mine meetings at the end of the day. After dinner he spent more time on the telephone, either receiving or making countless calls. A distinct change seemed to have come over him. He appeared to be deep in thought, restless, and more irritable than ever before. Momma asked us to let him have some space, to be quiet, and stay out of the living room while he was "handling business" on the phone. We knew something was wrong but we thought it would pass; it always had.

○ ✕ ○ ✕ ○

For my fifteenth birthday I got to pick the spot for supper, plus Daddy promised he would teach me how to drive. I chose the Sands Café in Rock Springs. Momma had baked my favorite devil's food cake with divinity frosting which we'd have for dessert when we got home. Packages from my family surrounded the cake in the middle of the dining room table. It started to snow as Daddy backed the car out of the driveway to go into Rock Springs.

As always, the Sands was packed with people. Momma, my brothers and I ordered shrimp fried rice. Daddy had the special—fried oysters. After we finished eating, Momma looked out the window and said, "Oh, my gosh. We'd better head for home, John. Looks like we're in for a big snowstorm."

As we left Rock Springs, the snow got heavier, making for near whiteout conditions. It was difficult for any of us to see the road ahead. "Gee! Look at all the ice on our windshield wipers. They can barely move," my brother Johnny exclaimed as he leaned over the front seat with his head practically on Daddy's shoulder. "Sit back in your seat, Son, so I can concentrate," Daddy warned. Everyone sat tensely, focused on the road through the headlights, not saying a word.

Even though Daddy had slowed way down, we could still feel our Hudson slipping on the icy, snow-covered highway. Suddenly, Daddy yelled out as he threw his arm across the top of the front seat to shield my brothers and me from being thrown forward, "Hold on! We're sliding!" Snow flew up on both sides of the car, and we could feel it careening down a steep embankment and finally coming to an abrupt stop. Daddy threw the car in reverse and gunned the engine trying to back up onto the highway but it didn't work. He couldn't get the car to move forward either. Exasperated, he threw both hands on the steering wheel and shouted, "God dammit, we're stuck!" He struggled to push open the door through the drift of snow on the driver's side. It finally swung open, and he trudged to the back of the car and opened the trunk. He grabbed an arm full of wool army blankets Momma kept in the car for emergencies. Like most Wyoming families, we'd learned to be prepared. Daddy said the tailpipe of the car was completely bogged down in snow. We bundled up in the blankets while he climbed back up to the highway to flag someone down for help. Within five minutes, without the car running, we started to feel freezing cold.

"Momma, I'm so cold my teeth are chattering," I sputtered out.

"Scoot over kids," she replied. Then before we knew it she was out the passenger door, back in the back door, and beside us. We cleaved to her under the blankets using our aggregate body heat to help stay warm.

Daddy, wearing only light clothing, dress shoes, and nothing on his head, made many attempts to claw his way up the embankment to the highway.

We sat in silence for what seemed like a long time before we heard voices. What a relief it was when Daddy threw open the car door and yelled, "Everyone okay in here? We got lucky. Someone stopped who can pull us out." His face was beet red, and it looked like his eyebrows and eyelashes were covered in frost. "Margaret! How the heck did you get in the back seat?" he asked as he slid behind the steering wheel.

"I managed to get out of the car the same way you did and open the back door just enough to squeeze through. I think we all would have frozen if I hadn't," she answered.

He told us to sit tight while the man hooked a chain to the frame of our car and pulled us out with his truck. Suddenly the car jolted and within moments we felt it slowly being pulled back up onto the highway. Staring out the back window, my brothers and I were jubilant watching the progress.

The good Samaritan removed his chain from our car and then tapped on the driver's side window. Daddy rolled down the window and, calling the man by his first name, thanked him for stopping and offered to pay him for his trouble.

"Oh, no. I don't need any money, Johnny. Glad to help. Just say it's payback for all the folks who have helped me at one time or another when I was stranded." He said he'd follow us back to Stansbury to be sure we made it.

Daddy shifted the car into gear, and we cautiously continued down the road. No one said a word as we traveled down the highway, anxious to see the lights of Stansbury. Johnny and Jimmy kept watching out the back window. Finally Jimmy announced, "That man is still following behind us, Daddy, just like he said he would."

"Who is that fellow, John? I don't believe I've ever seen him before. Thank God he stopped to help and didn't just drive by," Momma said.

"Someone you don't know, Margaret." After a pause, Daddy said, "I feel like two cents. Yesterday I had to fire that guy."

"Why, Daddy? Did he do something wrong?" I asked.

"It's a long story, and one I don't want to talk about and you don't need to know," was his reply. When we drove up into our driveway, the man honked his horn and flashed his headlights. Daddy stepped out of the car and waved back at him. When we opened the front door to our house, the deep warmth of coal heat engulfed us.

○ ✕ ○ ✕ ○

The storm passed, and we fell into our usual routines. How I wished I could invite Don to my house, as I did my other friends, but I was afraid it would upset Daddy.

Then, one of the neighborhood girls, not yet eighteen, ran off to marry "some cowboy" from out of town who was older than she was. It was the talk of the camp. Her parents were beside themselves with grief. The girl never came back.

When Daddy heard about this, he was livid. I overheard him tell Momma it seemed like one more example of local girls coming to bad ends. "I told you boys mean trouble, Margaret. If Marilyn ever did something like that I'd kill the damn kid and then kill myself."

"Oh, John!" she uttered impulsively with fear in her voice. "Don't talk like that. What's come over you? I don't want you to let the problems follow you home anymore. I never want to hear that crazy talk again! The kids hear that and what are they supposed to think? Now I mean it! No more of that crazy talk."

After that, it seemed like Daddy became even stricter, not with my brothers as much as with me. He watched every move I made. One hot afternoon my brothers and I, along with some of our friends, rode our bikes around camp. We girls all wore shorts and midriff blouses. I wore one of the outfits Aunt Ann had given me: tan short shorts with a yellow blouse.

Don and his friend had come over from Reliance on their bikes and rode around with us. Suddenly I heard Daddy's whistle, so familiar to everyone in camp, summoning us kids home. The whole group of kids headed for my house to see what was wrong. Momma and Daddy were out on the back porch.

"What's wrong, Daddy?" I asked as I rode my bike into the yard. Right there in front of all my friends he yelled for me to go into the house. I started to cry because I didn't know what I had done to make him so upset. When I got inside, Daddy confronted me, with Momma standing silently at his side. "I don't ever want to see you wearing those short shorts again. You hear me? Now go into your room and take them off, and then bring them to me so I can throw them away. I've told you a million times, I don't want you dressing like that!"

I ran into my room, slammed the door behind me, took off my shorts, put on my pedal pushers, and took the shorts out to Daddy. "Here! Throw them away. But as soon as I get eighteen, I'm leaving here to go and live with Aunt Ann and Uncle Zeke."

The next day at school, I told Don what had happened just because I was wearing shorts. He admitted that Daddy had scared even him and that he was afraid for me.

For the next couple weeks, I just stayed to myself whenever Daddy was home. Then one day clear out of the he blue, he apologized. "Those shorts were my favorite," I replied. "Aunt Ann didn't see anything wrong with them and neither did Momma."

Daddy said the reason he was so strict with me was because he loved me and that as long as he lived, he'd do everything in his power to keep me from making foolish mistakes. He said men in the mine talked, and he never wanted his daughter to be the topic of those conversations like some young girls in camp. I then realized that it was the girl who had run off with the cowboy that set Daddy off, and he took her actions out on me.

He then gave me an intense look and said, "And by the way, I've been asking around about that Don kid you've been seeing." My stomach dropped, and I waited for the bad news. "He comes from good stock. So from now on it will be okay for you to meet him at the dances and movies and invite him over to the house." My whole body relaxed when I heard him utter these words.

CHAPTER SEVENTEEN
DUST TO DUST

THE WINDS OF change blew into Stansbury. Dust fell on the empty streets and into the boarded up houses and eventually into empty rooms and hallways. Once beautiful yards dried up and became eyesores. The company quit painting the houses; if repairs were needed, the residents had to make them. Some people didn't, but most did. When a family moved, the house sat empty. Dust returned to its native home— dust to dust.

The families suffered the same neglect as the structures. Fathers cried at the kitchen tables: Where would they work now? Wives comforted the men through their own tears. Children cried themselves to sleep and then had nightmares at seeing their parents in such distress. No longer did company doctors come out to the camp clinic. Those needing attention had to go to the Miner's Clinic in Rock Springs.

Still people spent their weekends working in their yards, gathering weeds and debris that had blown in over the winter, fertilizing their lawns and tilling the garden soil for summer planting. Children worked alongside their parents shoveling all the trash into large cardboard boxes to take to the dump. Still holding onto hope, most people continued to maintain their houses and yards. Neighbors chatted over their fences as they worked, usually about the mine, the thing that held the town together.

○ x ○ x ○

In June, we again looked forward to the annual first aid contest. Even though many people had moved, surprisingly many teams entered. While UP had drastically curtailed its operations, it spared no expense in making this event special. We all celebrated just as we always had.

Afterwards Don and I went to the dance and couldn't hold each other close enough as we danced practically every single dance together. We were lost in the words of songs like "You Made Me Love You," "Hold Me," and "Secret Love," which we called "our songs." We danced in close embrace with my head on his shoulder and his cheek resting against mine. I thought how I was going to miss him when I went to Salt Lake during the summer. Tears came to my eyes as I whispered in his ear, "Please don't go with anyone else while I'm gone."

Right before midnight Daddy came walking toward us and said, "How about the last dance with your father." As we danced, I felt embarrassed because none of my friends ever danced with their fathers. All eyes seemed to be on us. Some boys started calling me "Daddy's Little Darling," which embarrassed me even further.

o x o x o

Sunday on the drive home after church, Daddy shocked us kids when he said that he had decided Momma needed a change of scenery. Since he was having to spend more and more time at the mine and in meetings, she was going to take the train with us kids when we went to Salt Lake. Then she'd continue on to California to spend a couple weeks with her sister Kay.

So the following Saturday we four boarded the train headed West. We had never known our parents to be apart before and worried how Daddy would fix his meals and pack his lunches with Momma gone. Jokingly he assured us he could survive for two weeks, but he looked forlorn standing alone on the depot platform as the train pulled away.

Uncle Zeke and Aunt Ann were waiting for all of us at the Salt Lake train depot. Momma visited with them a little while before she continued on the train to California. On the way to the house, Uncle Zeke and Aunt Ann told us of all the activities they had enrolled us in. When we pulled into their driveway, Jimmy leaned over the front seat of the car and shouted excitedly, "Whoa! Whose car is that?" There sat a beautiful yellow, Corvette convertible.

"Why, that's my car," Uncle Zeke replied proudly. "Your aunt has her Cadillac and now I have the Corvette. I thought I'd teach you kids to drive it around the block while you are here."

Jimmy sitting in Uncle Zeke's yellow Corvette in Salt Lake City. Once Jimmy sat behind the wheel of this car, I knew he never wanted to go back to the coal camp.

Jimmy could hardly contain himself, "I love that car, Uncle Zeke! "

"Well," Uncle Zeke smiled, "maybe someday it will be yours." Right then and there, when I heard Uncle Zeke's words and saw the look on my brother's face, I knew it would be hard for him to return to the life he had in Stansbury. He wanted that Corvette to be his.

We knew all the neighborhood kids from the summer before, and they were enthralled by our lives in a mining community "You mean your father works underground all day?" one girl asked in amazement.

I was surprised that most of the girls, all of Mormon faith, dreamed of getting married and having a large family rather than going on to college. To them, college was for boys who would be the providers of their families. I found this thinking strange.

One night while we were riding home, Aunt Ann said something that not only shocked me, but hurt my feelings as well.

"Too bad your father is a coal miner! You kids have missed out on so much of what life has to offer by living and going to school in those mining camps." My brothers and I looked at each other but didn't say a word. We'd never felt like we were deprived.

In the evenings Uncle Zeke taught Jimmy and me how to drive. Everywhere we drove, people stopped and stared at that Corvette. Soon we could shift the gears smoothly when we had our turn behind the wheel. "Now see, if you lived here, you could drive this car whenever you wanted after you got your licenses," he said.

Sometimes Aunt Ann took me downtown to shop in my favorite stores, ZCMI and Auerbachs, and to have lunch at the restaurant in the Hotel Utah. She bought me jewelry, shoes, and handbags to match the clothes Uncle Zeke continued to buy from the clothing salesman. She bought me fabric, patterns, yarn, and crochet thread by the case lot rather than by the skein or ball. "I know your mother likes to do this kind of handwork too, so there'll be plenty of thread and yarn for both of you," she explained as she paid.

I helped with house cleaning and sometimes ironed Uncle Zeke's shirts. Jimmy and Johnny mowed the lawn and took care of the yard.

○ ✕ ○ ✕ ○

One night after we the turned off the lights, Jimmy asked, "Do you want to go home when the summer is over, Marilyn?"

Shocked, I replied, "Jimmy, why would you ask such a question? This is just for the summer. We belong home with Momma and Daddy."

His reply took a while in coming. "I was just wondering."

Soon the day came when we had to go back home. As we gathered our things Aunt Ann looked like she was going to cry. I put my arms around both of them and thanked them again and again for making me the best dressed girl in high school.

Jimmy opened the car door and then ran over to Aunt Ann, clinging to her as tears ran down his cheeks. "I don't want to leave," he pleaded. She put her arms around him while she explained why he had to go back. Reluctantly he let go of her and slowly got into the car.

Momma and Daddy were waiting for us in Rock Springs. Momma looked so rested after spending two weeks in California.

Jimmy loved getting the mail out of the mailbox at Aunt Ann's house, partly because in Stansbury our mail was not delivered to the door.

The minute we stepped out of the car, Tippy began jumping all over us, licking our faces while his tail wagged back and forth like a whip. I could hardly wait to call Don whom I'd written to daily from Salt Lake.

At supper that night, plans were made to leave the next day for our family vacation at Granite. But this year, things would be a little different. Daddy had a chance to earn extra money by helping Uncle John, who owned Johnny's Bakery, install a new baking oven. He'd drive us up to Granite, set up the tent and campsite, and then drive back to Rock Springs. As soon as I could, I called Don to tell him that we were going to the mountains early the next morning.

The week passed quickly with swimming and fishing and hiking. For the first time in our experience, some Mormon families set up generators for electricity and lit up their whole camp area. We thought it was the greatest thing we'd ever seen.

○ ✕ ○ ✕ ○

The day after we came home our house was a madhouse with my brothers and me fighting for the use of the bathroom to get ready for school. After breakfast, we each gathered all our new school supplies, and Momma handed out our lunch tickets before we bolted out the back door. Johnny was so excited that he was the first one to get on the bus. But Jimmy held back and slowly climbed the steps into the bus. "Come on, Jimmy," I urged. "What's the matter with you? Don't you feel good?"

He didn't answer just looked up at me with a wry look in his eye. I knew Jimmy was having difficulty accepting going to school in Reliance. I leaned over toward him and said, "Jimmy, come on. Salt Lake is great for the summer, but our family lives here, and it's the law that you have to attend the school where your family lives."

"Marilyn, I hate this town and that dumb old school. I don't want to be here," he said and he began to cry. I felt sorry for him as I put my arm around him and begged, "Please don't say that. If Momma and Daddy heard you say that, they'd feel bad. They'd think you liked Aunt Ann and Uncle Zeke more than them. Look at all the good friends you have here—friends you grew up with. Tell you what. I have twenty dollars in my purse that I've been saving. If you'll try to be happy at school in Reliance, I'll give you the money."

"Promise?" he asked wide-eyed. "To spend however I want?"

"Promise. However you want." Excited about the money, a smile came across his face, and he hurried to be with his friends the minute we got off the bus.

○ ✕ ○ ✕ ○

After that first week of school, I called Jimmy into my bedroom and gave him the twenty. He looked down at the money and then up at me. "I can't take your money, Marilyn," he replied.

"No, go on and take it. That was the deal."

"No, I don't want it," he replied as he turned and walked back into the living room and sat next to Momma on the couch.

A of couple days later, Jimmy stayed home from school because he said he didn't feel good. I thought it was to avoid getting the shots that were scheduled that day. When Johnny and I got home after school and burst into the house, we didn't see Jimmy anywhere.

As we gathered around the supper table that night, Daddy asked, "Where's Jimmy?"

Momma looked rueful, took a deep breath, and replied, "John, I put Jimmy on a bus to Salt Lake this afternoon when I went into town for groceries."

"You did what?" Daddy bellowed, slamming his fist on the table.

"He just doesn't like living out here in the camps anymore. He said he'd run away if I didn't let him go back to Ann and Zeke's to start junior high." Her voice quivered. "John, they can give him so much more than we can. He hated the thought of going to Reliance school. I couldn't take listening to him anymore. So I called my sister and asked if he could live with them."

"You never even asked me," Daddy said incredulously.

"I knew you'd never agree, so I just put him on the bus while you were at work."

Johnny and I sat motionless, terrified of what Daddy's reaction would be. We didn't have to wait long.

"God dammit, Margaret!" he exploded, standing up from the table. His chair tilted back and crashed onto the floor. His face was sheet white, and I thought for a moment he was going to hit her. But he reached for his coat and headed for the garage. Without turning to look at us he yelled, "Okay, everybody in the car. Right now! We're going to Salt Lake to get your brother."

"But what about eating supper first? The kids are hungry." Momma stammered.

"To hell with supper! We're going to drive to Salt Lake right now and bring my son home!" Momma looked like a scared rabbit. We hurried to get our jackets and got into the car. He backed out of the garage so fast, I thought he was going to take down the garage wall.

Daddy drove fast along the highway leading out of Rock Springs not saying a word. About twenty miles later when we got just outside Green River, he pulled off onto the shoulder of the highway, lowered his head onto the steering wheel, then turned to look Momma straight in the eye and said, "You made the biggest mistake of your life today! One you'll live to regret." With tears in his eyes he choked out the next words, "I've lost my son because of you. Why in the hell did you do something like that?"

Momma took a Kleenex out of her purse and dabbed at her eyes as she looked out the passenger-side window. The silence weighed heavily over all of us. Daddy pulled the car back up onto the highway, turned around, and headed back home.

o x o x o

Daddy was never the same after that night. I wondered if Jimmy's leaving would cause my parents to get a divorce. After a week or so, he and Momma began talking again, and we kids thought he was back to his old self. But we all missed Jimmy, and we knew he wouldn't be coming home. There was a void everywhere around us, especially at the supper table or when Johnny had to sleep alone in their bedroom in the basement. "I don't want to sleep down there anymore, Momma. I get scared down there without Jimmy," he told her.

Momma never expressed missing Jimmy. It was as if his leaving were something she expected to happen long before it did. But, I knew she missed him. As for Johnny and me, Jimmy's defection was a terrible rending of our family's solidarity. All our lives, our family had strived to be a unit, to be special, different from the other families, and now we'd been torn apart.

We missed hearing Jimmy's voice, his infectious laugh, and his adventurous spirit. Momma always seemed to cleave to him the most. Now, she seemed to have a matter-of-fact acceptance of his leaving.

o x o x o

I noticed miners listening even more closely to the evening mine schedule report on KVRS radio. Radio announcers Imogene Parr and Michael Read, who were already household names for reporting daily mine schedules, became more important to the hourly miners than ever before.

I also started hearing talk about a new company that had begun operation outside of Rock Springs, that mined a hard rock I'd never heard of before—trona. I wondered what this new material was used for and why there was a demand for it.

CHAPTER EIGHTEEN

DARKNESS GATHERS

N OTICES POSTED in the *Daily Rocket-Miner* newspaper and in all schools throughout the mining communities announced the time and date of free polio vaccinations at the Old Timers Building in Rock Springs. Momma was scared of polio but at the same time she worried about letting us be inoculated with this new vaccine. Ever since the outbreak in 1953, she had read everything she could on polio and kept a file of all the articles. She remembered reading about the problems with the original Salk vaccine and that it actually caused 260 cases of poliomyelitis, resulting in ten deaths. She decided to talk to Dr. Muir; he assured her that the vaccine was safe and all children should be inoculated as soon as possible.

She waited until the day they were giving the shots before she told us we had to get one. I was scared to death and Johnny cried all the way into town and continued crying while we stood in the long line leading into the Old Timers' Building. Each of us stiffened and held tightly to Momma's hand when the nurse grabbed our arms and swabbed a small area with alcohol before injecting the needle. Tears even swelled in my eyes when it came my turn.

"Relax, now, and the shot won't hurt," the nurse said in a soft and soothing voice, as if we could. After the shot, we continued to sniffle pathetically especially when we overheard the nurse tell Momma and Daddy that we would have to have another shot in a month and a booster shot every five years. From that day forward, we were gripped by fear if we had a headache and a stiff neck at the same time. That, we thought, was a sure sign of polio.

When Momma realized how fanatical our behavior had become, she summoned the family together in the living room the evening after we'd gotten the shots to listen while she read the articles about polio she'd compiled.

Momma and Daddy were staunch Democrats who loved President Franklin Roosevelt; so we loved him too. His name was a household word, always invoked with the highest praise. The first article Momma pulled out of the folder was about Roosevelt's affliction with polio. This got our attention immediately. "In 1921," she read, "outbreaks of poliomyelitis plagued America. That summer, a young politician named Franklin Delano Roosevelt was vacationing with his family at their Campobello estate. After an exhausting day fighting a local forest fire, taking a cold swim for relief, and then lounging at home in his wet swimsuit, he went to bed feeling as though he had contracted a cold. A few days later, Roosevelt found out he had polio."

She read us details about how it is spread, the types of polio, the odds of becoming paralyzed, treatment options, the odds of recovering. We sat silently, horrified as we listened.

Momma then showed us pictures of children on crutches or in iron lungs. With curious eyes, Johnny looked up at Momma and asked, "Well how did that doctor discover how to make the medicine in the shots we took?"

"Oh, that Dr. Salk is a smart scientist," she replied as she leafed through her articles. "Listen to what Dr. Salk did, kids!" She pulled out another article and read the multi-syllable words aloud.

Johnny and I and even Daddy hung on to every word.

That same evening, Momma let us call Jimmy to be sure Aunt Ann had taken him to get his polio shot somewhere in Salt Lake. We each took our turn talking to him. It was so good to hear his voice again, and at the same time, it made us miss him more. The whole family wanted him to come home.

○ ✕ ○ ✕ ○

It felt strange for Johnny and me to catch the school bus each morning without Jimmy. I found myself wishing Daddy had followed through with his plans that night, and we'd driven to Salt Lake and brought

Jimmy straight home. It was still hard for me to understand Momma's actions. And I couldn't help but remember the many times last summer when Jimmy would remark, "We live like rich people when we're in Salt Lake, don't we, Marilyn?"

Did he miss hearing our voices? Sitting around the kitchen table? And how could he not miss all the friends he grew up with? And Tippy? Try as I might, I just couldn't figure out how he could walk away from all of this, plus Momma and Daddy, just for the love for money and material things. We had it so much better than some of our friends, with loving parents and a good, steady income. Was it dreaming of driving the Corvette that lured him away?

Oh, we talked to him on the phone, but that only made things worse. I could only imagine how terrible Daddy must have felt when those calls came. Looking back, even now I feel what Momma did was wrong, and from this perspective, I feel Jimmy acted like a spoiled brat. But maybe there were things I didn't know and never will.

One night while I was helping Momma do the dishes, Daddy came into the kitchen and said, "Come on, Marilyn. I'm going to teach you how to drive." I didn't have the heart to tell him that Uncle Zeke had already taught me. Excitedly, I ran to get Johnny who kept his mouth shut about me driving in Salt Lake. He crawled into the back seat, and I slid behind the steering wheel.

"You're really going to let her drive, Daddy?" Johnny blurted out.

"Daddy, make him sit back and be quiet so he doesn't make me nervous," I was quick to respond. I listened intently to his instructions as I slowly backed the car out of the driveway and onto the dirt street in front of our house.

Soon after that, I got to drive the family car, filled with my brother and as many friends as we could squeeze into the car, for fifteen minutes only, practically every night after supper as long as I didn't leave the nearly deserted Stansbury streets.

o x o x o

Our lives resumed the regular routine of the coal camp. Families gathered for pick-up games of baseball in the street, kids took excursions to climb the rock formations at Red Rocks, we went with Momma to watch movies at the community hall while Daddy went downstairs to the pool hall where he drank a cold beer while visiting with other miners. Don and I continued to see each other at school or when he drove over to Stansbury.

My freshman year was a joyous time at school. For the first time I was able to select some elective classes. My dreams of going to college, getting my degree, becoming a high school business teacher, and marrying Don now seemed like they were within reach.

I took as many subjects as my schedule would allow and strived to excel in them. Grades dictated who would be awarded the Valedictorian and Salutatorian Honor Scholarships. I simply had to be one of those recipients. When I entered junior high, Daddy told me he didn't like to see "B's" on my report card, and, for the most part, I didn't let him down. His approval still meant the world to me.

One evening after supper, I went outside to see if I could help Daddy put away the hoses and other lawn equipment for the winter. After we'd finished, he sat down on the front porch step and lit up a Lucky Strike cigarette. Daddy didn't smoke often but when he did, I loved the smell of the tobacco. "I can hardly wait until I'm old enough to smoke," I said.

With obvious irritation, he replied coldly as he looked down at me, "Women have no business smoking. There's nothing worse than seeing a woman with a cigarette hanging out of her mouth. So get that crazy idea out of your head right now. If I ever see you smoking a cigarette, I'll knock it out of your hand, no matter how old you are." And I knew he would.

○ ✕ ○ ✕ ○

Saturday morning Momma and I went into Rock Springs to give Grandma's house a thorough cleaning. We tried to do this at least twice a month. Grandma, whose health was failing, loved to have us help her, and when we were done she'd take a little black purse out of her pocket, take out folded money, and try to give it to Momma. "Ma, I don't want your money. I'm just glad I'm able to help you after all you've done for us."

As we drove away from Grandma's house, Momma suggested we get something to drink at the Lean Too Café. As always, the little café was packed, so Momma and I took seats at the counter. She ordered coffee, and I had a cherry coke. The smell of homemade pies filled the air when "Scoot," the waitress, took them out of the oven, her blond hair bobbing and her eyes sparkling. "Mmm, those pies sure smell good," Momma said. "If it wasn't so close to supper, Scoot, I'd have a piece."

After Scoot set the pies on wire racks to cool, she brushed some crumbs from her white apron, pulled up a stool on the opposite side of the counter, and visited with Momma. She intrigued me because she was a palm reader, and she always seemed to be right. "Your left hand is what you came into this world with, and the right hand is what you'll do with it," she would say in a serious tone before a reading.

"Scoot, will you tell our fortune today?" I begged, laying my hands out flat in front of her. She played with her big hoop earrings for a minute, and then almost resignedly said, "Sure."

People in the café looked over as she began, "Give me your hands and let's see what we have here," she replied. Softly she caressed my hands as she studied my palms: "The two lines right below your little finger means you will marry two times. I see two lines there."

"Two husbands?" I asked, with childlike naiveté.

Looking more closely, she continued. "That's what I see. Looks like you have a strong lifeline. See how it wraps around?" She stroked my palm. "You will go through a lot in your life. See all of these little spider lines all over your hand? You must beware, though, and never go around horses. Harm will come to you from a horse." Smiling she looked into my face and said, "Now frown for me," which I did immediately. "Looks like you'll have three children. You have three distinct lines running across the top of your forehead."

She placed my hand down and said, "Okay, now let me do your mother's." She took my mother's hand.

"Yep, looks like three children is all that you'll have, Margaret. I see you have a nice long lifeline as well, but I also see that you too will go through a lot in your life." Then a serious look came over Scoot's face. She paused for a moment before looking up into Momma's eyes and saying, "I see a lot of people coming to your house."

"Oh, Scoot, you're right. I'm having my family for Thanksgiving. My sister and her husband are bringing my son Jimmy home for the holiday," Momma answered in agreement. "You're pretty good if you saw all that in my palm."

Scoot's eyebrows pulled almost together in a frown and she continued, "I see a crowd of people at your house, but it won't be a happy crowd, Margaret."

"Well, I don't know why that would be," Momma said.

"Well, we'll just have to wait and see," Scoot said as she grabbed a damp cloth and wiped the top of the counter, signaling the end of the session. "I only tell what I see, and sometimes I am wrong, you know." She seemed eager to admit her fallibility.

Momma stood up to pay the ticket. "Thanks for telling our fortunes, Scoot. Marilyn just loves your fortune-telling."

A worried look crossed the face of the waitress as she turned toward the coffee pot. Then she turned back, and hugged Momma. "Take care, Margaret."

As we drove out to Stansbury I couldn't help but ask, "I wonder what Scoot meant? And, why wouldn't she tell you why she thought the crowd would be unhappy?"

"Oh, fortune-telling is just hocus-pocus. You can't take what Scoot says too seriously. Nothing but imagination and superstition. Scoot's only guessing. And, she'd have to be blind not to see how you hang onto her every word. She likes the way your eyes light up when she tells fortunes. You're a great audience."

Looking at Momma quizzically, I countered, "But Momma, Scoot's never wrong. People say she's like a gypsy when it comes to fortune-telling. I've never known Scoot to tell a fortune . . . that didn't come true."

CHAPTER NINETEEN

THE DAY
THE WHISTLE BLEW

CHRISTMAS CATALOGS began to arrive in the mail, and Momma was sprucing up the house for Thanksgiving. I just loved this time of year.

Miners continued to listen to the daily mine schedules on the radio each night to see which days the mine would be working or idle.

On Thursday, November 10, 1955, we woke up to a brutally cold morning. The wind chill had dropped the temperature well below zero. Every window in the house had been covered on the inside with heavy frost even before we went to bed.

At five that morning I heard Momma downstairs putting more coal into the furnace and monkey stove, so she'd have hot water for washing up and laundry. Next I heard her in the kitchen, making Daddy's breakfast and packing his lunch pail.

During the week, Daddy had voiced concerns to Momma about one area in the mine. He had reservations about the way the superintendent wanted it to be mined and was vocal about it at the mine meeting the night before. Today the miners would be working in that area.

I snoozed a few minutes longer as my parents talked over the breakfast table. Momma once again tried to talk him into buying the new suit he had tried on at Sinko's Men's Store in Rock Springs the weekend before. He insisted she should use that money for what we kids needed. "Oh, John," she replied, "you really need a new suit, and it wouldn't hurt you to buy something for yourself for once."

As he was getting ready to leave, he asked Momma to remind me not to cut articles out of the newspaper for my journalism class until he'd had a chance to read it.

I got out of bed, wrapped my robe around me, and leaned against my bedroom door. They didn't pay me any attention.

As Daddy reached for his winter jacket, he turned to her and said something I'd never heard him say before, "This is one day I wish I didn't have to go to work." Momma stood behind him and helped him pull his coat up over his shoulders. Then she wrapped her arms around his waist, laid her head against his back and said, "John, you've never missed a day of work in twenty years. If you don't feel like going in today, why don't you just stay home?" He slowly swung around and looked into her eyes, resting his hand on each side of her waist. "I can't do that, Margaret. The company is paying me."

After a quick goodbye kiss, she stood back and studied him, "Your hair looks so gray this morning, John. Guess we're both getting older. Both of us are starting to get more gray hair."

"Well, got to go, Mumsy." He called her by the pet name he sometimes used. "My men are waiting for me. Oh, did Ginger Kovach call about that Westvaco job?" He tucked the thermos under his left arm and grabbed his lunch bucket. "I talked to him Monday night and he said he would get back to me sometime this week. You know, Margaret, as soon as I get his call, I'm quitting this damn job and going to work in the trona mines."

"No, he didn't call, or I would have told you right away. I know you're waiting for that call," Momma replied. As she walked him to the back door, she confessed that since it was close to payday, all she had for his lunch was crackers and cheese, a slice of apple pie, and two apples.

"That's okay. Probably won't have much time for lunch, anyway," he said, without any real concern, and he went out the door.

After Daddy left the house, Momma went to my bedroom window and scraped a hole in the frost, so she could watch him walk down the street. She tousled my hair when I joined her at the window.

Every morning Tippy came out of his doghouse and waited at the bottom of the steps, eager to walk alongside Daddy on his way to the mine bathhouse. Then Tippy trotted back home and waited for my brother and me to come outside so he could walk us to the school bus. At the end of the day, he'd be waiting at the school bus stop to walk us home. Then he'd go to the mine office to walk Daddy home.

"Sure you don't want to go back into that doghouse this morning, boy?" we heard Daddy ask as he knelt down to pat Tippy on the head. Wagging his tail, the dog jumped up and down excitedly, attempting to lick Daddy's face. "Well, come on then, boy," Daddy said. "Let's go." From out of the darkness Slim Nielsen, Daddy's best friend, stepped under the light of the street lamp, and together they walked down the snowpacked road to work.

○ × ○ × ○

I knew the routine at the bathhouse from hearing about it for years. Daddy always seemed taken aback by the miners' deferential greeting when he walked into a room. The bathhouse was filled with the morning crews, changing into their work clothes, putting on their mine hats with the carbide lamps, hanging the methane detectors on their belts, and then stopping in the bathhouse office to grab brass employee ID tags off the scheduling board. Every man had one of the tags, and it was a rule that whenever they went down into the mine, it had to be in their left shirt pocket. Daddy reached for his tag, number 701, slipped it into his pocket and sat down to read the morning fire boss report depicting the condition of the mine upon early morning inspection. Then in the cold, dark morning, the miners trudged through the deep snow over to the mine portal where the mantrip waited to take them underground.

The ride down underground was usually noisy, with men trying to talk to each other above the sounds of the mantrip clinking its way over the steel rails. Yet, some days the men were silent. Today Daddy's crew was unusually quiet because they knew the danger of the mined-out seam where they'd be working. As the story goes, one man hollered out, "Johnny, how long do you think we'll be in there?" Daddy yelled back, "Just long enough to clean it up and get the hell out."

When they got to the number seven seam, the mantrip stopped and the crew unloaded. The silence in number seven room was unnatural until one miner yelled, "Hey, Johnny! Did the fire boss report say this place was okay today, nothing unusual, no movement or anything?"

"His report was positive," Daddy yelled back. "Let's get going."

Once underground, Louis Julius, a 38-year-old machine operator, headed for the Joy Loader at the face of the seam. From pictures and

The entrance to the mine looked insignificant and didn't give an indication of the miles of tunnels underneath the ground.

newsreels, this machine always reminded me of a giant miner, lying on his stomach, arms outstretched, pulling the coal toward him, first one arm and then the other. And the machine did this real fast, while at the same time loading the gathered coal onto a conveyor belt which led from the machine to coal cars behind. "Hey Johnny, I'll back the loader out whenever you're ready," Julius yelled climbing up onto the loader to start the engine.

Just before one o'clock that afternoon the cleanup in the work area was almost done. Newspapers later reported that Daddy hollered back to John Maffoni, an older, experienced miner, and asked Maffoni to swap lunch time with him so Daddy could finish up. Maffoni headed out of the seam leaving Daddy, Louis Julius, and George Chenchar at the face. Daddy hollered for Julius and Chenchar to get out while he backed the loader out. Julius yelled over the roar of the loader engine that it would take longer to switch places, so he'd just back it out. But first, he asked, should one more roof bolt be put in?

"No!" Daddy replied in staccato. "Move out! Now!"

At that very instant, without any warning, all hell broke loose. A low rumble was followed by an enormous underground thunder, everything went dark, and the whole ceiling crashed down as the three remaining miners made a frantic dive for cover under the Joy Loader. After the rock had settled, Julius, face down under the rock in a muffled voice faintly called out, "Johnny? Johnny? Johnny, are you there? Are you all right."

After a while Daddy answered in a slow, distressed tone, "It's my back, Julius. It's my back. Holler as loud as you can." Instantly, a second cave-in came as tight as it could fall. All movement and sound ended as abruptly as it had begun. An absolute quiet filled the site, which was every bit as loud as the cave-in.

Julius called out for help again and then heard what were Daddy's last words: "I'm a goner, Julius."

Digging operations began immediately with three rescue crews sent into the mine, alternating in the work. The work was slow in the cave-in area as miners had to work around the crumbled roof, which had caved all the way from the surface about 75 to 100 feet above, 2,500 linear feet from the mine portal.

O X O X O

Back at the house, Momma turned off the washing machine and went upstairs to make herself a bowl of soup for lunch before continuing with the laundry. She glanced at the kitchen clock hanging on the wall to see what time it was and realized that the clock had stopped exactly at one o'clock. She'd have Daddy fix it when he got home. She stepped outside to feed the dog and noticed he was in his doghouse, and she couldn't get him to come out to eat his food.

O X O X O

At one o'clock that afternoon, I was sitting in study hall where everyone was talking about the newest James Dean movie, *Rebel Without a Cause*, that would open at the Rialto Theater that weekend. I had just opened my bookkeeping book and was starting to do my homework when Mr.

Chadey, the superintendent, burst into the classroom with an ashen look on his face and nervously made the following announcement: "There has been an accident at the Stansbury Mine."

The whole room was instantly quiet. Stunned, I felt like all the blood had drained from my body. This was the first time we kids had ever heard about a mine accident while we were at school, so I knew it had to be bad. Silently I wondered who was injured and why Mr. Chadey looked so strange. But before I had a chance to think any further he continued, "The following students are to come with me immediately: Don Nichols and Marilyn Nesbit." I froze. My thoughts reeled, my name echoed in my head. There are no words to describe the sickening feeling that comes over you when you learn there has been a terrible accident and you are the one being singled out to go to it. A tumbling kaleidoscope of fears immediately filled my mind.

My whole body went numb. I was almost unable to move, speak, or comprehend what Mr. Chadey had just said. Surely there had to be a mistake. My thoughts jumped back to the day I overheard Daddy talking to Momma when he came home after a bad mine accident: "Margaret, if I'm ever hurt bad in a mine accident, I would rather be dead than have to spend the rest of my life in a wheelchair." Those words resounded over and over in my mind until I covered my ears trying to stop them.

There had been many mine accidents before, but they always involved other people, never my father. He was the boss, the one who helped rescue other miners. I reasoned that surely this was the case now, and Mr. Chadey had only called my name so that I could go along to comfort Don Nichols, an upperclassman also from Stansbury whose father was less experienced than Daddy.

We both started gathering our books when Mr. Chadey commanded, "No, don't take your books. We don't have time for that. Let's get going!"

Don and I followed him out of the room and ran for our lockers to get our coats, as Mr. Chadey kept saying over and over as we ran to catch him, "Hurry. We've got to hurry." The more he said it, the more scared I became. I had never seen Mr. Chadey so distressed before.

When we got outside, the freezing weather almost took my breath away as Don and I jumped into the back seat of Mr. Chadey's car.

Someone had left the engine running so it was warm inside. "Has my father been hurt, Mr. Chadey?" I asked, as we sped down the highway leading to Stansbury. Looking at his reflection in the rearview mirror, I saw his face become chalk-white when I asked that question. For a while he didn't respond, just continued looking straight ahead at the road in front of us. Then he muttered in a voice I could barely hear, "Things are still pretty sketchy. But, Don, I think your father is one of the ones injured."

"Oh, no!" I started crying for Don and threw my arms around him trying to console him. Don just sat there with a scared look on his face. "Don't worry, please don't worry," I pleaded. "My father used to be a safety inspector. He knows what to do when there is an accident, and he's down there with your dad." Don didn't respond as the tears ran down his cheeks.

When we reached Stansbury, the mine whistle was blowing incessantly from its tall pole outside the mine office. Never in my life had I seen so many people, trucks, and cars at the mine. Instead of driving toward the mine, where everyone else was, Mr. Chadey drove directly to my house and then took Don to his house. I never said a word as I shoved open the car door, jumped out, and ran directly inside my house into Momma's arms. "Oh, Momma, I am so scared! Has something happened to Daddy? Oh, Momma! Oh, Momma! Say he's okay! Please say he's okay!" I wailed as I buried my head into her chest.

Momma didn't answer, but as she held me close I felt her body trembling. I threw my head back to look up at her. The look on her face put a deep fear in my heart. I'd never seen her look so distraught, so lost, so afraid. Tears poured down her cheeks, and she couldn't speak. "No, Momma, don't cry! Please. I get so scared when you cry," I wailed.

Then I noticed my Uncle John, Momma's oldest brother standing in the room, and I hadn't even seen him when I came in. I knew something terrible must have happened because even though he was my mother's brother, he had never, ever come to our house. Without saying a word to either of us, he suddenly turned and walked out the door.

"Why is he leaving, Momma? What did he say to you to make you cry so hard?"

She lowered her head and said, "I was downstairs in the basement, doing the laundry, and the next thing I knew, he was standing at the top of the stairs looking down at me. He just walked into the house, didn't even knock, so he startled me, you know, when I saw him standing there, and then he said, 'Well, Nesbit got his today.'"

"What did he mean?" I asked dumbfounded.

"Marilyn, as soon as Johnny gets home, we have to go down to the mine. I don't know what happened, and I'm afraid to go see for myself, but I don't think your dad made it today. There was a cave-in in number seven. The rescuers are down in the mine right now trying to get to the men who were under the fall. They think your dad is one of those men."

"They *think*, Momma! Think? Who *knows* for sure until we hear for ourselves? I just know Daddy will be okay and that Uncle John was wrong. He never gets things straight. Hurry, Momma, let's go down to the mine."

"No! We have to wait for your little brother. Mr. Chadey forgot to pick him up. He should be here any minute. I don't want him to come home to an empty house. We can't go until he is with us." With her voice quivering she looked deep into my eyes and said, "Oh, Marilyn, pray! Pray harder than you ever have before that your father is okay. It must be so dark and dusty down there right now that it would be hard to identify anyone. That's it. They just don't know where everyone is."

Things were happening so fast, we didn't have time to think straight. The phone started ringing incessantly as we waited for my little brother. Calls were coming in from relatives from the area as well as those from out of state who had heard of the accident on the radio. All Momma could say as she talked hurriedly was, yes, there had been a cave-in, Johnny was down there, they'd better come, and without waiting for a reply, she'd hang up. Then she called Aunt Jenny, Daddy's sister who lived in Rock Springs. "Jenny, please get out here fast. I need you! I think John has been hurt in a mine cave-in!"

I could hear Jenny's screams. "Oh my God, no! Bob and I will leave right now," she exclaimed.

My brother Jimmy frantically called from Salt Lake. "Momma, was Daddy in that cave-in? Is he going to be all right? Is he, Momma?"

Then Aunt Ann took the phone from him and said they would leave Salt Lake immediately.

After Momma finished talking to Aunt Ann, she went into the bathroom and closed the door. I stood outside the door waiting for her to come out. I didn't know what else to do. After a couple of minutes when I didn't hear the toilet flush, I knocked on the door and asked, "Are you okay, Momma?" There was no answer. Then I yelled and banged on the door, "Momma! Are you okay? Please come out of there right now! Don't leave me alone! You're scaring me!"

The bathroom door slowly opened, and Momma stepped out and threw her arms around me. Sobbing hysterically she said, "Oh, Marilyn! I just about did something very stupid. But your banging on the door made me get ahold of myself."

"What were you going to do?" Things were happening too fast for me to comprehend.

"For awhile there, I thought I just couldn't do this. I don't want to face any of this. I was looking for something in the medicine cabinet to drink to make all this stop. I wanted to die."

"Oh, my God, Momma. You can't do something crazy like that. We need you! Daddy needs you! He's going to make it out of there, Momma. I know he will!" As we started to put on our winter coats and scarves, Johnny came bursting through the front door yelling, "What's wrong, Momma? Why did they take me out of school? Why doesn't the mine whistle stop blowing?"

Momma threw open her arms to him sobbing, "Oh, my baby! Come here. Your daddy has been in a terrible accident. I have been waiting for you to come home from school so we could all go down to the mine. Mr. Chadey brought your sister home a few minutes ago but in all the excitement, he forgot to pick you up. "

"But is he okay, Momma?" Johnny asked over and over and then started to cry when he saw Momma crying. After a few moments, her crying subsided and a strange calm came over her. As if transformed, she stood up. "Come on kids," she motioned while putting her arms around both of us. "Now that we are all together, we have to go down to the mine right now! But get really bundled up because it's a blizzard out there!"

Down the snowpacked road we ran toward the mine portal, Momma gripping our hands like a tourniquet, as we braced ourselves from the freezing wind and blowing snow. Our legs couldn't keep up with hers. When we started to slip on the ice or fell, she yanked us up, screaming at us to keep up with her. Johnny sobbed from the pain of being jostled. I could feel my heart beating like a drum in my chest. Momma ignored everything and kept running.

As we got closer to the mine, people were coming from everywhere, flowing like blood through an artery. Cars and trucks, filled with women, kids, dogs, and older people, slowed down and shouted for us to get into their cars, but Momma didn't seem to hear or care. We ran the whole way. I could hear dogs howling in the distance, along with the sound of Momma's labored breathing and the frightened sobs of my little brother. I could feel the blood pounding in my head as I ran to near exhaustion, no longer feeling the ground beneath my feet. I was running so fast, I felt like I was floating.

Crowds of people surrounded the mine for as far as I could see in any direction through the wintery whiteout, making it difficult to get close to the portal. Some walked around with uncertain purpose, their eyes searching, lips mouthing names they couldn't bring themselves to say out loud yet. The rest stood stiff and silent, like roses in winter. Momma pushed forward determinedly, as if she knew a destination that was worth arriving at.

I stayed focused on the mine portal and despite all the commotion going on around me, it was hard to believe anything bad had happened inside that portal. It didn't look any different than it always had. Even though it was below zero, miners and their families stood in protective huddles, attuned to the environment, listening to the arcane sounds around them, waiting for news of recovery from UP officials. Overriding it all was a pervasive feeling of organized chaos, frenetic movements, shrill sounds, and dozens of unexpected crises that all had to be attended to at once. Everyone seemed to be caught up in an endless, terrifying nightmare, waiting under intense pressure for what was to come.

I looked off in the distance and saw a steady procession of cars making its way up the highway; I could hear engine noises and saw smoking tailpipes and headlamp lights clear down the road as people came to witness the worst mine accident at the Stansbury Mine.

○ × ○ × ○

The mine whistle was deafening. I wanted it to stop. Rescuers came out of the mine covered with coal dust and blood. The lamps on their hard-hats bobbed and fluttered like fireflies. Even though it was freezing, no one seemed to notice the cold. My brother and I stood in awe of what was unfolding around us. A sick feeling started in the pit of my stomach as the soft weight of my little bother's hand sneaked into mine. We overheard men saying that all the miners had been ordered out of the mine immediately after the cave-in. It was so severe that dust from way down below ground came billowing out of the portal. Even the land above ground, over the cave-in site, had subsided. A command center was set up at the mine office, down the road from the portal. Martha Tarufelli and Gloria Fabiny drove to their homes to brew as much coffee as they could, packed the containers in the trunk of their car and offered coffee in paper cups. No one left to go home for supper that night.

Ambulances parked near the mine portal, and we could hear women sobbing and sometimes screaming as injured miners from the vicinity of the caved-in area were brought out of the mine. People swarmed around Momma who was bundled in her only wool coat with a wool scarf tied around her head. In the hurry to assure we kids were warm, she had forgotten her gloves and her hands were red and shaky as she rubbed them together incessantly, both from nervousness and to keep them from freezing. She looked so fragile as she strived to endure the weather, so like a fawn lost from its mother. Some men, whom we recognized as mine officials by their dress suits and long wool coats, approached Momma and took her aside. Talking to her in low tones, she listened intently to what they said with her head down, cupping her freezing hands over her mouth. Johnny and I stood like wooden soldiers until I couldn't take it any more. I simply had to get closer to the mine portal.

I tightened my grip on my brother's hand and together we ran, pushing our way through the crowd, headed straight for the mine entrance, both of us asking anyone we recognized along the way, "Have you seen our dad?" But no one answered. I thought everyone was treating us rudely, like we were just kids and had no business near the mine. Finally, my brother and I stopped. We were as close as we could get to the portal. We just stood there, looking deep into the dark mine watching for Daddy to come walking out, refusing to believe that he wouldn't. Far below the ground, the men who once shared lunches while complaining about the company or had told tall tales about women were working feverishly to rescue or be rescued. Miners never give up on a man lost in the mine until he is brought up, dead or alive.

The darkness grew thick, and heavy snow blanketed everything. Johnny and I stood together motionless, shivering underneath our winter garb until finally I had to do something to keep us both from freezing. I opened my coat and pulled him next to me, so I could wrap my coat around him and share our body heat. I felt his little body shaking against mine. The wind blew right through the pretty coat I had to have, even though Daddy had said it wouldn't be warm enough for winter. I began to shiver convulsively. All the while, rescuers continued going in and out of the mine, gathering in knots of grave discussion where I overheard them say, "Johnny's down there." They had so much blood on them, they looked like they'd been in a train wreck, which only heightened my fears.

The gloomy minutes turned into hours. The howling wind took on an bone-chilling sound and dropped the temperature even further. My little brother looked up at me with his big brown eyes now red and swollen and quietly asked, "Are you scared, Marilyn?"

"No," I replied feigning confidence. "It won't be long now until we see Daddy, and we can go home." The flagrant falsehood seemed to re-assure both of us for a moment. But deep down, for the first time in my life, I was terrified. Though we were standing in a huge crowd of people we'd known all our lives, we felt completely alone.

Suddenly, I heard Momma call to us from inside a parked car which I recognized as belonging to Malcolm Condi, a local United Mine Workers official. "Marilyn, come over here right now. We've been looking everywhere for you kids. We've got to stay together right now! Do you hear me? Don't go running off into this crowd again. They'll let us know when there is any news." Then her voice softened to a near whisper. "You poor kids look like you're freezing."

Quietly I started to pray, "Please, oh please, dear Lord, let Daddy be okay. We need a miracle. Please don't let him be so badly hurt that he has to be in a wheelchair."

Slowly the mantrip could be seen coming up out of the mine, and I glimpsed some miners hurrying to meet it. Without saying a word, they worked with intense, caring precision to lift an injured miner out of the mine car onto the ambulance stretcher.

A wife, who was summoned by a rescuer, ran to her injured husband as he was loaded into the ambulance. Hugging him over his filthy clothes, cupping his blackened face, she kissed him again and again, so happy that he was alive.

"Look over there, Momma," I shouted. "They just brought someone out of the mine."

"Stay right where you are, kids." Momma instructed sternly.

But in the confusion of the night, I disobeyed her. "Johnny, you stay here with Momma," I said as I jumped out of the car and ran toward the ambulance

"They say it's Claude Nichols," I overheard someone say.

"But where is my father?" I screamed above the whistle and confusion. "He was working with Mr. Nichols." Still, no one answered me.

Then I saw Dr. Muir, our family doctor, ready to get into the mantrip to go down into the mine—something I had never known any doctor to do. Seeing him there, I just knew everything would be okay, for why would they send a doctor into the mine if someone was already dead? I ran to his side, threw my arms around him and buried my head in his chest. "Oh, Dr. Muir," I sobbed, "do you think my daddy is okay?"

Dr. Muir stood tall and straight, wrapping his long arms around me, saying nothing but looking up into the sky. I saw a terrible sadness that I had never seen in him before. He knelt down in front of me like Daddy used to crouch down to my level to explain things to me when I was small. He clutched me by my shoulders. I started a humming cry, and crying was the last thing I ever wanted to do in front of him. "You be strong for your mother, Marilyn," he said looking deep into my eyes.

"Why?" I screamed the word at him. Before he could answer, a rescue miner yelled loudly, "Doctor, we have to go! Now!"

He stood up, bent his head down to kiss me on the forehead, turned and ran toward the mantrip without looking back. He jumped aboard one of the cars already in motion, filled with rescue miners going back down. I watched the mantrip until it disappeared from sight, hearing only the rumbling of steel wheels over track leading deep into the mine.

Only later did I understand that Dr. Muir had risked his own life by just going into the mine and then crawling through a tunnel which the miners had made by digging with their bare, bloodied hands, handing rock and coal, piece by piece, to the person behind them in a human chain. It was an extremely dangerous area and yet Dr. Muir stayed down there, hoping the rescuers would find the miners who were suspected to be under the fall. It was beyond dangerous, as the removal of just one piece of coal could bring the whole ceiling down on everyone.

I never saw Dr. Muir when he came out of the mine, but learned later he had talked to reporters. He said when he reached the area of the cave-in, he saw men—many of whom had been down there all day—crying, covered in blood and coal dust, working feverishly to rescue the trapped miners, fearing yet another cave-in and being trapped themselves. He reiterated that once he reached the scene he saw "a lot of brave and courageous men working in adverse circumstances, men who made me very proud to be a member of the human race."

I felt the soft touch of a hand on my shoulder, and I turned around to see Aunt Jenny, Daddy's sister. She wrapped her arms around me and led me back to the car where Momma waited. Aunt Jenny was the sister who'd encouraged Daddy to come to Wyoming so many years ago.

Just then there was a knock on the car window. It was my friend Geraldine who said some kids from school were nearby in a car—did I want to sit with them for awhile? "Five minutes and then come right back," Momma said.

I walked over to their parked car, hoping maybe my boyfriend was inside. When I opened the door, cigarette smoke came pouring out. Don sat in the back seat with a group of kids Daddy had forbidden me to have anything to do with, kids I thought were lucky because their parents weren't strict. I had always thought they were popular, and sometimes yearned to be associated with them, and yet I couldn't bring myself to get into this car. Don said he couldn't get his car started, so he had hitched a ride with them, hoping to find me. I made excuses as to why I couldn't get into the car and quickly returned to where Momma, Aunt Jenny, and my brother were waiting.

Malcom Condi, who had gone down to the mine earlier, slowly came walking back to the car, and asked Momma and Aunt Jenny to step out. He put his arms around both of them and was talking in such low tones that it was impossible to understand what he was saying. Almost from the moment he started to speak, Momma and Jenny laid their heads together, put their hands over their mouths and started crying, but this time it was a deep, hopeless, helpless, hurting cry. I caught the glint of the wedding ring on Momma's frozen hand. She seemed to be losing her strength to stand, gripping Mr. Condi's arm like a vice. "Marilyn, Johnny!" she cried.

"Momma, what's wrong? What did he tell you?" I asked. Seeing the look on Momma's face scared me and my little brother so badly that we went from crying to screaming over and over again, "What did he tell you?"

Sobbing so hard she had difficulty getting the words to cross her lips, she said, "Daddy didn't make it this time, kids. Your daddy just didn't make it."

"No, no!" I screamed. My whole body ached, my head felt like it was going to burst apart. Even though I was crying so hard, not a sound was coming out of my mouth. Suddenly I felt someone grab my arm and as I turned around, my brother Jimmy was standing in front of me. "Oh, Jimmy! You're home; you came home. I'm so glad to see you! You should never have gone to live in Salt Lake. Daddy hated it. He never was the same after that. Please don't go away again."

"We came as fast as we could," he replied.

"Come with me, Jimmy, closer to the portal so we can watch for them to bring out Daddy."

"Marilyn, he's not coming out this time." The serious look on his face jolted me into reality more than anything else had. "Momma wants me to get you away from here so we can go home. Come on, Marilyn, let's go home," he said as he extended his hand to me.

"I'll go, but I won't believe he's dead until I see him with my own eyes. God can't take him from us, Jimmy, he just can't."

○ × ○ × ○

The rest of that night was a blur. I remember that we were like disembodied souls, all of us. I didn't even want to go inside our house. Surprisingly the lights were on inside the house, and Aunt Ann and Uncle Zeke's Cadillac was parked in front. What was once a lovely, loving home was now an empty shell that was ghostly quiet. Bill Lewis met us at the front door. "I kept the furnace going for you and the kids, Margaret. It's below zero out there tonight, and I didn't want you to come home to a cold house. Hope you don't mind me letting myself in. Your sister and her husband just got here from Salt Lake."

"Oh, Bill, thank heavens you thought to do that. I'm so tired I could drop. Thank you."

The first thing I noticed when we walked into the kitchen was Daddy's empty chair at the kitchen table. Traces of him were everywhere we looked: his towel and toilet articles in the bathroom, his side of the bed smooth and unwrinkled, his clothes in the closet, his black onyx ring and gold dress watch on top of the bureau, one of his coats hanging on a hook in the back hallway, and all his tools in the garage. Absent was the sound of his voice, his laugh that filled the entire house the minute he came home from work, and the way he whistled.

Aunt Ann was busy making places for everyone to sleep. "You and Uncle Zeke can sleep in my room, Aunt Ann, because all of us kids are going to sleep in Momma's bed," I said.

"Why, you can't all fit into that bed," she said in a sympathetic voice.

"Yes, we will," I replied. She didn't question any further.

Johnny broke the silence when he asked, "Where's Tippy, Momma?"

"I don't know," she replied distractedly.

Johnny ran out the back door, then returned carrying the big dog in his arms.

"Momma," he exclaimed. "He's been in his doghouse all day. His fur is freezing," he said, wrapping his arms around Tippy.

"Oh, my God, even the dog knew something was wrong! He never stays in his doghouse all day," Momma murmured. "Go sit by the furnace heat vent with him, Son, and rub your hands through his fur to help him warm up quicker." Jimmy ran to get Tippy's food and water dish which was on the back landing.

Finally we all went to bed in Momma's bed, with Tippy lying on the floor at the foot of the bed.

But we couldn't fall asleep knowing Daddy was still down in that cold mine. "Momma, I can't sleep. I want to go back down by the mine and wait until they bring him out," I said.

"Marilyn, it is freezing out there, and it won't do any good for you to stand out there. You have got to stay in bed. None of us will get any sleep tonight, but we have to try to get some rest while we wait. That is all we can do now."

"But Momma," I pleaded, talking low so not to keep my aunt and uncle awake, "I won't believe he's dead—until I see him."

"Oh, kids, I don't know when they will be bringing him out. The cave-in was so bad, it may take a while to get to him. But when they do, I don't want you kids to see him when they bring him out of the mine. Now I mean this! I don't want you to see him. Remember him like he was," Mother said.

"But Momma!" I wailed, "It's so cold and wet down there. Let's go back to the mine and find Slim. If anyone knows where Daddy is, it will be Slim."

Momma sighed, "Slim hasn't left the mine since the accident. He's down there right now working with the rescue teams to find your dad, and I know he won't come out until he is able to bring your dad with him. There is nothing we can do but wait."

During the night, I heard someone knock at the door. Momma got up and put on her robe, instructing us to stay in bed and not to come out. We scurried to listen at the bedroom door. Ann and Zeke got up and went into the living room. They all spoke in low tones, but I overheard one of the voices say they might not be able to bring my father out because of the danger to the rescue workers; they might have to seal off the seam. I could not contain myself and ran almost hysterically into the front room. "No, no, no! You can't let him stay down there; you just can't let him stay there. He would never do that to you. Make them bring him out, Momma."

Mother was frantic. We were now at the mercy of the UP mine officials, who left as abruptly as they came. "Go back to the bedroom kids, " Momma instructed. "So help me God, I will not let them leave your father down in that mine. But, we can't do anything about it now. We have to wait until morning before we can do anything about it."

Somewhat assured and not knowing what else to do, we climbed back into bed. Momma sat up for a while talking to Ann and Zeke, but when she finally came to bed, we all tried to put our arms across her. Throughout the night my brothers and I took turns sleeping where Daddy had slept just the night before. The fragrance of his Old Spice aftershave lingered on his pillow. What would we do without him?

○ × ○ × ○

As the sun came up the next morning, friends and relatives began arriving at our home. Thank goodness Aunt Ann was there because Momma was in no shape to greet anyone. Grandma and Grandpa were first to arrive. They had gone to California to visit Aunt Kay and had no sooner arrived than they got the news. Aunt Kay and Uncle Charles had brought them right back home, driving straight through the night.

Momma started crying all over again when she saw them. "Ah, Dad! What am I going to do?" she pleaded.

"You a strong woman, Margaret. The Lord is the only one who can help you through this now," he consoled her in his broken English as he held his lovely daughter in his arms. "Your Mother and I will always be here for you. You know that. Now brush away your tears. It isn't good for the kids to see you crying." Momma seemed to get comfort from her father's words and having her parents with her.

Johnny and I found some comfort in having Jimmy home. The three of us being together made it seem like the family was right again. Yet just when we'd start to feel the easy familiarity of family, reality would hit us, and we'd remember what brought us together.

Members of Momma's homemakers' club brought in delicious meals and beverages on this morning and every day for the next two weeks. They brought so much food that it was difficult to find a place to put it. Lola Nielson, one of Momma's best friends, took control of the kitchen. The women served the food and cleaned up the kitchen while their husbands visited with the men in the living room, all going over and over the accident. Momma couldn't thank them enough for all their work and generosity in feeding a lot of people.

Uncle George, Aunt Edna, and their four boys flew in from Riverton. Uncle George was working on a rig in the oil fields the night Edna heard the news on the radio. They packed up and left for Stansbury the minute he got off the midnight shift.

Daddy's sisters and brothers, who had also heard about the accident on the radio, flew in from Indiana and stayed at Jenny's house in Rock Springs. None of them had ever been to Wyoming before and came to Stansbury immediately upon their arrival.

All night long, dimly lit lights were on in every house in camp, with weary families unable to sleep, wondering when a tragedy like this would happen to them and when Daddy would be brought out of the mine.

Desperate to know exactly what had happened, Momma stepped out onto the porch to get the newspaper. Everyone gathered in the front room to listen as Uncle Zeke read the article on the front page. In it was a statement made by the only one who knew for sure, the man who was right down there with Daddy, Louis Julius. The *Rocket-Miner* had interviewed Louis Julius from his hospital bed:

Johnny told us to move out. "Let's get the hell out of here!" We were going to have three timbers across the face. The roof was good. We had one safety timber up. We were going to put up two more, so they could plug it.

There was no warning. The roof behind me fell, an enormous underground thunder, and I started to run. Some rocks were falling in front of me. I ran into something and fell. Johnny was a foot right behind me. I was knocked part way under the loading boom of the Joy loader. I crawled a little farther in. I asked Johnny how he was. He complained of his back and said, "I'm a goner, Julius."

Then the second cave-in came. My legs were buried to the knees, I was lying on my stomach. I called to Johnny again. But he didn't answer anymore, just moaned once or twice. I didn't think there would be any more cave-ins after that second time. It had fallen as tight as it was going to. Then all movement and all sound ended as abruptly as it had begun. An absolute quiet filled the site which was every bit as loud as the cave-in.

The Rock Springs Rocket-Miner *reported the cave-in and the aftermath.*

I saw a pant leg and reached down for it. It was mine. I started to dig with my hands to get the circulation going. It was cold, but I didn't have any trouble breathing. I was conscious all the time. Got dizzy a couple of times, but I'd start moving around, digging with my hands, trying to move my legs. That kept me warm.

I realized I still had my goggles on. I took the damn things off and threw them aside. There was a mine hat lying a little ways from me. I kept looking at it and wondering if it was mine. I asked them when they came to dig me out. They said it was Johnny's.

No one said a word after Uncle Zeke finished the article. The silence was broken by a knock at the door. Two officials from the mine wanted to talk to Momma in private. "Whatever you have to tell me can be told to everyone in this room. We are all family. But give me just a minute," she said as she stood and walked over to my brothers and me. "Marilyn, we're about to run out of coffee. I want you to take your brothers and go down to the store and buy another can of Folgers coffee and come right back."

"We want to stay and listen to what they have to tell you, Momma," I replied. "Then we'll go."

"Don't you understand, I want to offer them a cup of coffee, and we are all out. They will be here when you get back. So come on now, get your coats on and get going."

Jimmy and I put on our coats and hats while Momma bundled Johnny up against the frigid weather. We headed for the UP store with Tippy coming along. We ran all the way. After buying the coffee, we started back home. Just as we rounded the corner of the building, Tippy darted out onto the highway and began running behind a long, black vehicle that had just entered camp headed in the direction of the mine. It was as if Tippy knew.

My brothers and I knew what kind of vehicle this was and ran in the same direction until we reached the embankment on the side of the portal. The mantrip came slowly out of the mine, pulling empty cars, except for the front car which carried a blanket-covered stretcher followed by two cars filled with weary rescue miners who had been in the mine all night. When they saw my brothers and me standing there all alone on the snow-covered embankment, they ducked their heads and looked away.

Our eyes focused on the stretcher covered with a gray army blanket. For a few minutes, we stood motionless, as did the men after they exited the mantrip. They then proceeded to lift the stretcher in unison off the front car, setting it carefully on the ground. They stood next to it with their arms at their sides, heads lowered, and wept. The cold wind eerily swirled the snow around the scene.

Breaking the deafening silence, Tippy lunged forward and ran toward the stretcher, tugging at the blanket and making whining, whimpering sounds. One of the miners moved quickly to grab him by his collar and pull him back. Then the miner knelt beside Tippy and began stroking the anguished dog's back while softly commanding. "No, Tippy! No boy. Stay."

Moving as if in slow motion, engulfed in tears, my brothers and I were drawn toward the stretcher. When we reached it, we knelt down beside it. The miner holding Tippy's collar brought him over to where we were kneeling. For a few moments, none of the miners said a word, nor made any attempt to pull us away. Then, overcoming our revulsion and hoping against hope, my brother Johnny reached for the blanket covering the body on the stretcher.

"Hold it, kids!" a miner said as he walked quickly to where we were kneeling and knelt to gather us in his arms, hugging all three of us close, his coal-dust-covered jacket rough against our damp cheeks.

Then we recognized the voice as Slim's. Looking up at him with pleading, red eyes, my little brother asked, "Slim, is that really our daddy?"

Slim gasped and looked toward the heavens. "Ah, come on, kids. You don't want to see him like this. Come on now. Let me take you home to your mother."

I was fifteen, Jimmy thirteen, and Johnny was only ten years old. God only knows what the boys were thinking. I knew the man on the stretcher was our father, and he wouldn't be coming home to take care of us anymore. What was going to happen to us?

As we walked away with Slim, we took one final look back at the man on the stretcher who meant the world to us. The man who had worked more than twenty years in the mine he loved so much, never missing a day of work. I was always so proud when he came walking out of the mine with his men at the end of the day. He was handsome even when covered with black coal dust. I thought he was the smartest, strongest, most invincible man there was.

Now he was gone. He would never see us grow up, marry, and have children of our own. We had spent endless hours at the mine, riding a wave of emotions of raised hopes and grimly dashed dreams of a miracle rescue. Now, we were forced to face reality. Like the mine, our world had caved in around us.

When we got home that day, and I set the can of coffee on the kitchen counter. I looked over at Momma who was sitting with Grandma and Grandpa on the couch. There was an peculiar sense of remoteness in her eyes, a look that said she had retreated deep inside herself to a different place, a different time. The living room was filled with relatives and some ladies from the homemakers' club. No one spoke. We kids took off our coats and went to sit as close as we could to Momma. Johnny sat on the floor between her legs with his head on her lap, Jimmy and I sat at her sides with our heads in her lap also. She had her arms around all three of us running her fingers through our hair. We didn't have to ask what the company men had said. We knew the wait was over. Daddy was out of the mine. A part of me went with him that day.

○ × ○ × ○

A couple of days later, the wind and snow blew even harder, like wild horses galloping across the Red Desert with nothing to stop them across the long, snow-covered prairies. Momma sent me and my brothers to the post office for the first time since the accident. Among all the cards and letters was a letter postmarked November 10, 1955, the very day of the accident. The return address was Westvaco, Green River Works, Green River, Wyoming. Momma read the letter and said, "The trona company offered your father a job working in the new mine, the very day he was killed."

Early one morning a couple of days before the funeral, a car pulled into our driveway. Louis Julius, the survivor of the accident, stepped out of his car. We all met him at the door. He held his cap in his hands and said, "I came as soon as I was released from the hospital, Margaret. I came to tell you what Johnny wanted me to tell you and the kids." He was a tall, thin man whose face was all bruised, and he had a black eye. Momma, Jimmy, Johnny, and I gathered around him. We hung onto his every word as he told us exactly what it was like when the cave-in occurred and what a great person and boss he thought Daddy was.

"Did he say anything? Tell us exactly what he said, his last words," Momma begged.

Mr. Julius lowered his head, "There wasn't much time, and I could barely hear him. But I did make this out: 'Margaret, my kids, love.'"

Then we all cried, even Mr. Julius. Our whole world had been shattered in an instant. Daddy was gone and, for whatever reason, our prayers had not been answered.

Daddy, wearing his hardhat and mine clothes, before heading to work at the Winton mine.

CHAPTER TWENTY
SAYING GOODBYE

THE WEATHER seemed to go from bad to worse from the day of the accident until after we buried Daddy. My brothers and I compared the weather the day of Daddy's funeral to that in the Biblical story right after they crucified Christ, except instead of fierce thunderstorms blackening the day, the winter blizzard created a whiteout. We were terrified about what was going to happen to us. Who would help us, where would we get money to live, and where would we go when we had to move out of the company house?

Friends and relatives filled our house to capacity day and night, yet we felt all alone.

The wind blew so hard that at times it seemed like our little house could not withstand all its fury. The falling snow that heaped in huge drifts throughout the camp and the freezing temperatures made going outdoors treacherous.

◇ ✕ ◇ ✕ ◇

Momma needed to go into Rock Springs to the J. Warren Opie Funeral Home, and she insisted my brothers and I go with her. The funeral director, Mr. Opie, was a tall, stately man whom we had known all our lives. But even though we knew him, the fact that he was an undertaker intimidated us. The minute we opened the door, we noticed that distinctive smell of a mortuary and stiffened. Calmly Momma looked down at us and said, "Come on now, kids. You've got to be strong. This is something we have to do." We could see Mr. Opie waiting for us in his office. Hand in hand, Jimmy, Johnny, Momma, and I slowly walked through the parlor, paused, and approached his office apprehensively.

He somberly stood and walked over to greet us. "Margaret," he began, as he encircled all of us with his long, outstretched arms, "I can't begin to tell you how sorry I am that you have to be here today." My brothers and I had a difficult time holding our composure when he said this, but knew we had to be strong like Momma.

We kids sat impatiently in the chairs around Mr. Opie's desk, waiting for Momma to fill out the paperwork. The eerie feeling heightened. As we looked around, the dark, red velvet chairs, drapes, and carpeting seemed to close in on us like a cocoon. We didn't utter a word unless asked a direct question by Momma or by Mr. Opie, who was trying very hard to make polite conversation.

Finally, Momma laid the pen on top of the documents, stood up and said, "Come on, kids. Follow Mr. Opie and let's pick out a casket for your father." We followed them into a room filled with caskets of all different designs and sizes. We kids clutched.

"I don't know where I stand with money right now, Opie," Momma admitted, "so could you point out some of your less expensive caskets?" As we walked through the aisles, he pointed out the features and cost of each one. Suddenly, I could not remain silent anymore. "Momma," I whimpered as tears swelled in my eyes, "get him one that is air tight so the bugs can't get to him . . . no matter what it costs."

Everyone fell silent until finally Mr. Opie said, "Kids, just pick out the one you want. Don't worry about price. I'll charge the lowest price for whichever one you choose." His words made the decision less difficult. Johnny lingered in front of one particular casket. "I think we should get this one because it has a nice cross inside for Daddy to look up at."

"Then that will be the one we'll take, Son." Momma seemed relieved that one of us had an opinion.

While we kids returned to the office, Momma went outside to the car. She reappeared carrying a large box. Carefully she opened it, quietly explaining to Mr. Opie, "These are the clothes we want him to be buried in." Running her hand slowly across the fabrics, she continued, "He has worn this blue suit for many years, but it is the only suit he has. I tried and tried to get him to buy a new one, but he just didn't like spending money on himself."

"Which reminds me," Mr. Opie interjected, "what do you want done with the work clothes he came in with?"

Startled, Momma paused, then replied, "Just have them laundered and give them to someone who could use them."

Mr. Opie then reached into a drawer and handed Momma the two things Daddy had in the pocket of the work shirt he had been wearing: a metal measuring tape and his brass ID tag #701, dramatically bent.

"And one more question, Margaret. Who do you want for pallbearers?" Mr. Opie asked. Momma lowered her head and quietly named Daddy's dearest friends: Slim Nielson, Raino Matson, Bill Lewis, Nick Kragovich, Gerald Neal, and Charles Hanley.

Mr. Opie assured all of us that he would do everything he could to make the funeral nice. "Will he look just like he always did, Mr. Opie? Will you wash his face so there isn't any coal dust on it?" Johnny asked looking up at him with hopeful eyes.

"He'll look just like he always did, Son," Mr. Opie replied and then continued on, "The viewing will be tomorrow night." He then gave each of us kids a hug as we thanked him and headed out into the bitter cold.

On the drive back to Stansbury, Momma was silent. I couldn't tell if she was lost in thought or intently focused on the icy highway leading to our house, a house none of us really wanted to go back into. But as empty as it now had become, it was our only safe haven.

Finally Momma said, "I don't know how I am going to pay that bill."

After a long silence, I asked, "How much was it, Momma?" She explained the casket, vault, and headstone cost $2,300 and that included a five-space plot at the cemetery. But, she said, she couldn't worry about it right now. Too many other things were being thrown her way.

When we reached Stansbury, it was good to see the lights on inside our house. While we were at the funeral home, even more relatives had congregated in our living room. Members of Momma's Busy Bee Homemakers' Club were busy handling all the food people continued to bring. There was so much food, they ran out of storage space so Momma asked that they share the food with other families in camp so things didn't spoil.

O X O X O

The next evening we returned to the mortuary for the viewing and saw Daddy for the first time since the accident. For the first hour, only family and relatives were allowed in the sanctuary. We clutched each other when we saw him lying in the casket with coal dust still on his scalp under his wavy black hair, and the crying began all over again. We had to keep touching him, kissing his cheek, laying our hands on his. After our family's private time, a long procession of people from the community passed by his casket to say their final farewells.

After several hours, Momma began urging that it was time to leave, but we didn't seem able to go. Finally, Mr. Opie came over to us and said, "Come on, kids. Let me buy you a Coke over at the café across the street." He was finally able to coax us to leave.

O X O X O

On November 15, the following Tuesday, five days after the accident, we held Daddy's funeral at the Methodist Church in Rock Springs. The weather did not let up. While everyone was getting dressed, I went to Momma's room and asked, "Do you think anyone will be at the church today with the weather like it is?"

"We'll just have to wait and see," she replied. "It's frigid out there."

Momma drove our car to the funeral home, then we rode in the mortuary's family car, following the hearse to the church. The roads into Rock Springs were a sheet of ice, and the ground blizzards were so bad we could hardly see the road. We were all afraid that we would have to turn back. Out of desperation, Momma nervously prayed aloud, "Oh, dear Lord, show us the way." Miraculously, the sun broke through the blinding snow just enough for us to see the road ahead. I was so proud of Momma and the strength she exhibited. I don't know what we would have done if she had faltered, even once.

O X O X O

When we arrived at the church, the entire church parking lot, as well as all the streets for blocks around the church, was filled with cars. People stood all bundled up outside the church because there was no more room inside. Loudspeakers had been set up outside the church so the mourners outside could hear the service. Scoot Jolly was standing right inside the entrance of the church. She grabbed onto my hand and squeezed it as I walked by. So much had happened since the day she had told Momma's fortune—which turned out to be so true—it seemed like it was decades ago. Once we were inside the church, it was difficult for us to make our way through the crowd lining the aisles to the family seating area.

The church was filled with flowers and plants. The minister later told us there was over three hundred arrangements, more than he had ever seen for a funeral.

Daddy's favorite hymn, the one he really loved to sing himself, *The Old Rugged Cross,* began to flow through the room. For a moment I felt like this wasn't a funeral at all, that it was just another Sunday church service. I kept waiting to hear Daddy's voice as I had so many times before. But the reality quickly returned.

Lucille Smith, the soloist, stood to sing *In the Garden.* She and her husband were friends of my parents. She started crying so hard she finally had to step down, unable to finish the hymn. The Reverend David Rose delivered a beautiful eulogy, which made us even prouder of Daddy when he related what a compassionate person and boss he had been to the three hundred men who worked under him.

During the funeral, I turned my head to glance back at the crowd. My eyes suddenly met those of Dr. Muir who had gone into the mine the night of the accident. Our eyes met, and we seemed to lock into each other's thoughts. I remembered the day he told Daddy, "You get her out of here as soon as she graduates from high school, John." At the time, I didn't realize what he meant.

When the service was over and everyone had filed out of the church, Momma and we kids went forward to see Daddy one last time before they closed the casket. This time when I noticed how the coal dust was still on his scalp, it seemed right. He looked very peaceful. Johnny reached into his pocket and took out three small photos he had brought from our album at home. I was amazed he thought to do this. They were of us, of course, and there was even a picture of Tippy. He lifted Daddy's hand and slid the pictures under it. Jimmy wanted Mr. Opie to open the bottom lid of the casket to see if Daddy had his shoes on.

Mr. Opie quietly explained that no one is buried wearing shoes, an explanation Jimmy seemed to understand.

o x o x o

The cemetery was engulfed in a whiteout, and the temperature was below zero. As heavy snow continued to fall, ground blizzards swirled throughout the cemetery. I wondered how the groundskeepers could have dug the hole for the casket with the ground so frozen. The mortuary had set up a tent-like structure for our family to sit beneath, but the miners and their families all had to stand unprotected.

For as far as I could see from atop the hill stood Daddy's fellow miners and their families. Some miners wore their hardhats with the lamps on. The lights from the lamps wavered through the blowing snow like apparitions. Tears froze on the cheeks of men and women alike. I was overcome by a feeling of deep sadness, wondering how we could go on and how Momma would be able to take care of us. I had a feeling of foreboding about our lives from this point forward.

o x o x o

After the burial, the family gathered at Uncle John and Aunt Mildred's for a huge dinner prepared by Momma's homemakers' club. Uncle John met us at the door. Momma's first comment was, "I couldn't believe how many people were at the funeral, standing out there in that freezing weather. I wonder what John would have said if he could have seen all those people."

Uncle John, who always liked to joke, responded like the stiff snap of a ruler on a desk, "Oh, how the hell could he see. He's dead!" I looked at him in shock the minute he uttered those words.

Momma looked at him with real anger, hurt, and disappointment. For the first time she turned on him, and he looked startled by the vehemence in her tone, "What kind of a brother are you? Why did you even bother to come out to the house the day of the accident?"

"The only reason I was there was because Opie called me and asked me to go," was his quick retort. Why he acted as he did, I never understood. Maybe it was his way of showing grief. In any event, he broke Momma's heart when she needed her brother the most.

After everything was completed and the last person said goodbye,

Momma drove us back home. When we reached our house, Momma turned the engine off, and we just sat quietly crying for the longest time, not saying a word. None of us wanted to get out of the car. Finally, Momma broke the silence. "Come on, kids. This has been a long, hard day. We have to go into the house sooner or later."

As if in a trance, we opened the car doors and followed her up the steps. It was freezing cold inside the house. We'd been gone all day, so the fire had gone out in the furnace. It was so cold in the house that our teeth chattered as we followed her downstairs to start another fire in the furnace. Johnny shoveled coal out of the bin to help.

We missed Daddy even more at that moment and longed to hear his voice. Then I remembered how we could hear his voice again. Several years earlier Daddy and his friends had been in a local saloon where a man was making recordings of miners singing. Daddy had a beautiful voice and loved to sing his favorite Scottish songs in his Scottish burr, so his voice was recorded. I ran to my bedroom and dug through my cedar chest drawer until I found the 78 RPM record.

We left on our coats and hats as we gathered around my phonograph and listened to the songs he had recorded so many years before. The following stanza of him singing *A Cottage for Sale* seemed to fit:

> *The key's in the mailbox the same as before,*
> *But no one is waiting for me anymore,*
> *The end of our story is told on the door*
> *A Cottage for Sale.*

After playing the record several times, all three of us kids took off our coats and shoes and once again climbed into Momma's double bed with her. This time we left our clothes on to keep warm. There was barely enough room, but we didn't care because we just wanted to be close. I finally fell asleep reliving the events after the accident. Then it struck me. Of all the people who came to pay their respects at our home, the mortuary, or the funeral, the superintendent who had sent Daddy to get the last bit of coal out of the ill-fated number seven seam was not one of them. He didn't even send flowers. I went back to sleep hating that man and blaming him for what happened.

○ ✕ ○ ✕ ○

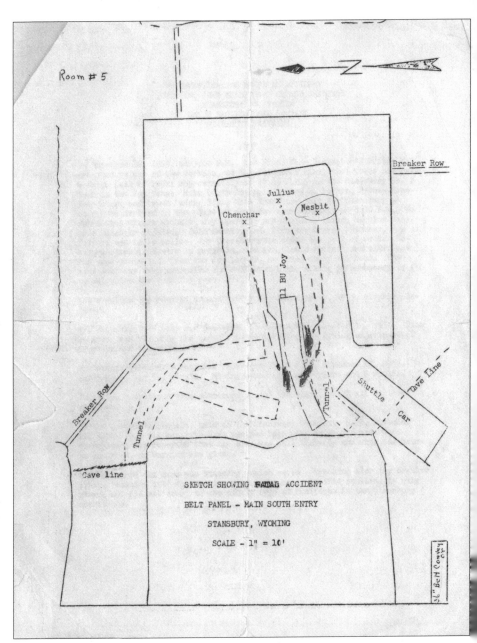

The drawing from the State Of Wyoming investigation into the accident showing the layout of the mine shaft.

On November 16, one day after we buried Daddy, the State of Wyoming held a court hearing investigating the fatal accident. Bill Villanova, Sweetwater County Coroner, opened the court held in the Stansbury Community Hall. Momma did not attend. Even though I was only fifteen years old, I wanted to go, but Momma would not let me. The company probably wouldn't have allowed me in the door anyway.

A panel of jurors from the Stansbury Mine was sworn in: Charles Hanley, Joe Zaversnik, and Mike Begovich, Jr. Also in attendance were State Bureau of Mines officials, the State Mine Inspector, the Deputy Mine Inspector of the State Mining Department, and representatives of Union Pacific Coal Company, including the Stansbury Mine Superintendent. All of these men knew my father well, both personally and professionally.

As I read and reread the entire transcript of testimony, I kept in mind which agency each questioner represented. The jury made the final recommendations:

(1) That the mined out area be watched closely, if caving does not take place, that caving be induced by blasting the roof, stumps or both.

(2) That all timber be placed in breaker rows, be good timber and placed close together and in a workman like manner.

(3) That only one panel stump be mined at one time; finishing one stump, then start mining on another. That the remaining stumps be mined with a shaker conveyor.

When I saw the third recommendation, I knew why Daddy hated to go into the mine the day of the accident and why he'd disagreed with the mine superintendent about the way to mine the number seven seam. The shaker conveyor method would have been slower than using the Joy Loader, but safer for the miners. Maybe this is why he disagreed so vehemently with the superintendent whose concern seemed to be getting the coal out as quickly as possible.

Only two people knew what really happened in number seven seam, the superintendent who sent him there and Daddy. Thus, I'd never know the truth.

This photograph from another mine shows the process of putting up a roof bolt. (www.miningusa.com, provided by Dave Zegeer)

CHAPTER TWENTY-ONE

A DOSE OF REALITY

"IT WASN'T supposed to be like this. It's always too soon to lose someone you love especially when you don't have a chance to say goodbye," Momma agonized again and again. Everyone asked her the same questions, and when they didn't, she asked them of herself. "Where are we going to live now that we have to move out of the company house?" "How can I raise my children alone and hold us together?" "How am I going to have enough money?"

Since the day of the accident, a continual flow of people had come to the house offering assistance. One of the most memorable visits was the morning Daddy's best friend, Slim, and his wife, Lola, came over to see how we were doing. He brought all the clothes and personal effects Daddy had stored in his wire basket in the bathhouse, along with Daddy's lunch bucket.

Momma gasped when Slim handed her the bucket, "Oh, my God, Slim," she said as she pulled off the lid. "All I had in the house that morning for his lunch was crackers and cheese. He didn't even get a chance to eat lunch. I'm so embarrassed opening it now."

We gathered around, curious about what Daddy kept in the bathhouse basket. Slim placed on the table the clothes Daddy wore to work that day, a nylon comb, and a new box of Lifebuoy soap. I remembered how Slim and Daddy had walked to work together that day and how the next morning they'd finally ridden out of the mine together on the same mantrip.

"Why did this have to happen?" Momma asked as she laid her head against his chest, and he wrapped his arms around her. Slim stood quietly and then said in his deep voice, "I wish I knew. But one thing I do know is when the good Lord has his hand on your shoulder, Margaret, there's nothing anyone can do. It was just his time, that's the only way I can figure."

He stepped back, reached into his coat pocket and pulled out a large, bulky envelope, and handed it to Momma. The envelope looked like it had passed through many coal-dust-covered hands. Momma started to cry again when she saw all the money stuffed inside: ones, fives, tens, twenties, and fifties. Slim explained that the money came unsolicited from miners who wanted to give her a helping hand. Momma asked him to thank all of them and to tell them their money was a godsend. Slim and Lola continued for years to be a support system in our lives.

○ × ○ × ○

Later that same afternoon, Daddy's brother, who also worked in the Stansbury mine and lived in the house behind ours, came over. Uncle Bob was completely different from Daddy. We rarely interacted with him or his family. It seemed like he was always wanting something for nothing and was jealous of his own brother. So it didn't surprise me when he asked if he could have Daddy's dress clothes and shoes.

"No," I shrieked with a disdainful look at Uncle Bob. "Don't give Daddy's things away now, Momma. We only just buried him." Momma, who hadn't even thought about the clothes, agreed it was too soon. She told him we needed time to go through Daddy's things.

○ × ○ × ○

Aunt Ann and Uncle Zeke were the last of the relatives to go home. After the funeral, they'd stayed in Rock Springs with Grandpa and Grandma to give us some time and space. A day before they left, they brought Grandma and Grandpa out to see us. Realizing our tenuous financial situation, they asked if Jimmy and I could both live with them until Momma got back on her feet.

"But I don't want to go to school in Salt Lake, and I don't want to leave Momma and Johnny," I replied when Aunt Ann came into my bedroom and told me what she had in mind. I hurt her feelings, and she snapped, "Well, if you don't come with us, I'll take back all the clothes and things that I have given you."

"Take 'em!" I snapped back. "I am not going."

She then went out to where Jimmy was sitting in the living room and asked him, "Are you coming back with us?"

"No, Aunt Ann," he answered. "I can't. I have to stay here now. Momma needs me."

This resulted in a lot of arguing with Aunt Ann, and all of us said things that we wouldn't have if we hadn't been under so much strain. Aunt Ann painted a bleak picture of what was in store for us if Momma didn't let Jimmy and me live with them. At one point she even went into my bedroom and started taking clothes out of my closet and stacking them on my bed to carry through with her threat.

After she did this, I looked in my closet and saw only the five dresses Momma and Daddy had bought me hanging in my closet. I threw myself across all the clothes on my bed and sobbed. Grandpa came into my room, sat down on the edge of my bed, and gently stroked my head. "Don't take it so hard, Mala." he said in a calm tone. "Bad for your blood pressure. Your aunt is upset. She doesn't mean what she is saying right now. Everything has a way of working out. Be patient with her. This is a stressful time for everyone."

The whole house was in an uproar. Grandma and Grandpa were getting upset, and I was afraid Momma would buckle under to Aunt Ann's demands. After things calmed down and the room got quiet, I said, "I know that you have our best interest at heart, Aunt Ann, but we can't go with you. We belong here with Momma. I'll box up all the clothes you gave me, so you can take them back. I can get by with what I have. But I just can't leave Momma and my brothers. And Jimmy can't go either. He never should have gone to live with you in the first place. It broke Daddy's heart. You don't know how terrible that made him feel. It was like you stole his son."

"I never stole his son," she retorted in amazement. "He wanted to come live with us. What kind of life did he have in this coal camp?" she questioned, her face ashen.

Suddenly, seeing the hurt look on my face, she hesitated, then said, "Oh, Marilyn, I don't want your clothes. I shouldn't have yanked all those dresses out of your closet like that. I'm just beside myself with worry about all of you living out here alone in this mining camp. I could never live in a place like this. Your mother hasn't worked for twenty years, and now she is going to have to be the breadwinner. How will you get by? And where will she find work? The only jobs for women around here are waiting tables or taking in other people's washing and ironing. I'm just so worried about you all. We thought it would be easier for her to get back on her feet if we took two of you kids off her hands."

My face burned from frustration. I felt exhausted, spent from crying and emotion. I said very calmly, "Aunt Ann, I know you mean well. Momma says that all the time. But she's all we have left now, and we need to be here with her. I'm worried, too. But I'd worry even more if I was away from Momma. Besides, I want to be able to visit Daddy's grave at the cemetery. Just give us some time to work things out. We can visit you and Uncle Zeke—if you'll let us—but we need to live here with our mother for now."

Soon after that, Grandma and Grandpa left with Aunt Ann and Uncle Zeke. I went to my bedroom window and watched their car go down the snow-covered street wondering what they must be thinking as they drove out of Stansbury. When I thought of all the things she'd predicted might happen, it scared me. But at the same time, I knew we kids could not leave Momma no matter what we had to go through. We were where we belonged, and this is what Daddy would have wanted for his family.

○ ✕ ○ ✕ ○

On Monday, November 20, ten days after the accident, Johnny and I went back to school. But Jimmy wanted to stay home with Momma that day as he still did not want to go to Reliance school. In fact he refused to go there. Momma was beside herself as to how to handle him.

After spending that first summer in Salt Lake, he never again liked living in Stansbury, and he was very vocal about this. He called Uncle John and Aunt Mildred in Rock Springs and begged them to let him live with them so he could go to school in Rock Springs. Mildred and Johnny had a big house with plenty of extra room. If he lived there he could still see Momma, but he wouldn't have to live in the camp.

We never learned what he said to them. All we knew was that John and Mildred agreed to let him live with them if Momma would pay the $100 the school district charged for an out-of-district student. Grandpa gave Momma the money, and she enrolled Jimmy in Rock Springs school. He agreed to work after school and on weekends in Uncle John's bakery to pay for his room and board. It was a sad day when we packed up his belongings and drove him to Uncle John's house. My brother was only thirteen years old, and he was leaving our family again. Now it would be just the three of us in Stansbury.

Johnny and I were apprehensive about returning to school after the funeral. The bus was unusually quiet on the way to Reliance. From time to time, I'd catch our friends staring at us. None of them knew what to say. The kids at school reacted the same way that first day back at school. None of our friends had experienced the death of a parent, so they were at a loss as to how to show their sympathy. Daddy was the first miner that they knew who had been killed in a mine accident.

I walked Johnny from the bus to his classroom in the grade school. When he saw his favorite teacher standing at the door, he anxiously blurted out, "I'm sorry I missed so much school because my dad got killed." She knelt and wrapped her arms around him.

I walked slowly toward the high school where Don was waiting for me. He'd come to our house a few times during the days after the accident, but I was always surrounded by other people. Now, here we were, alone for a few minutes before the bell rang for our first class. The look on his face told me that he knew the depth of my sorrow. I looked at him and said, "Don, I don't have a father anymore." He took my hand and together we walked into the building.

He stayed close to me that entire day, which began with a National School Assembly. Sitting on the bleachers beside him, I couldn't concentrate. I sat there thinking about all that had happened in my life in a few short days and how I was different from everyone else in that auditorium because they all had fathers. Not wanting to draw further attention to myself, I fought back tears. I was already tired of people staring at me remorsefully. I longed to be my old self again.

○ ✕ ○ ✕ ○

A few nights later, we went through Daddy's clothes, one item at a time. When I saw the blue shirt we kids had bought him for his birthday, I recalled when he'd joked, "Kids, thanks for the nice shirt, but please don't buy me any more blue shirts. Blue shirts always make me feel blue!" He didn't have many things in his closet and somehow we recounted, with occasional input from Momma, when and where he had acquired each item, right down to his neckties. When we finished, one whole side of the closet was empty. Momma quickly spread the hangers holding her clothes across the closet to fill the void.

On the following weekend, Momma called Uncle Bob to come over and pick up the carefully packed boxes of clothes he had asked for. When he got there, he asked if he could also have Daddy's tools and the camping gear because as he put it, "You guys won't be going camping anymore." Momma, looked shocked, but she agreed to let him have the things. It made me heartsick when I heard his words for I knew he was probably right, we wouldn't be going to Granite anymore, and we never did.

Bob made many trips up and down the basement stairs, each time coming up with his arms loaded with boxes to put in his car. Johnny and I sat on the back porch watching him, never offering to give him a hand and anxious for him to leave. We suspected once he took everything he wanted, we'd never see much of him or his family again, and we didn't.

That evening when Momma went downstairs to put more coal in the furnace, she noticed how empty the basement was. Besides the tools and camping gear, Uncle Bob had taken all the cans of paint, painting supplies, and lumber Daddy had stored in the basement. When she came upstairs, she said, "Well, kids, it looks like Bob really cleaned us out."

○ ✕ ○ ✕ ○

It's funny how things happen in life. Right after Daddy was killed, two popular songs came out which contained some lyrics every one of our friends thought summed up Daddy's life in the mine, as if the songs had been written about him. We thought of Daddy each time we heard Jimmy Dean sing *Big John* and Tennessee Ernie Ford sing *Sixteen Tons* on the radio.

○ ✕ ○ ✕ ○

Living through the accident was devastating, but little did any of us know just how rough the days ahead were going to be.

Don drove over every school night. Johnny would join us at the kitchen table, and we'd all do our homework. Whenever our house needed something fixed, Don was right there to assist. He was a good handyman. Momma treated him like a member of our family, and he provided some of the male presence we missed so much.

○ × ○ × ○

One afternoon Johnny visited at Grandma's house in Rock Springs while Momma and I picked up a few groceries. On our way back to Grandma's house, I spotted the Lean Too Café and asked, "Can we stop and get a cup of hot chocolate?"

"Oh, that does sound good on this cold, wintery day. I think I have just enough change," she replied, and we pulled up in front of the restaurant.

Once inside, we sat at the counter and Scoot, the palm-reading waitress, came to take our order. Momma and Scoot visited while we sipped our hot chocolate. Before we left, I turned to Scoot and asked, "Did you really foresee the accident the day you read Momma's palm?"

With a shocked look on her face, she replied simply, "Everything is written in the palm."

As we walked toward our car, I couldn't help but notice how pretty Momma was and how nice she looked in her long, grey wool coat. I was so proud to be her daughter and so admired the strength and resolve she showed. I looked up into the clouds and felt Daddy's presence, like he was looking down at us with a smile on his face.

After Daddy's accident, Momma was confronted with continuous challenges that she had never even considered previously.

CHAPTER TWENTY-TWO
PICKING UP THE PIECES

WE KIDS HAD always looked forward to the holidays, but this year was different. Momma was continually confronted by important issues she had to resolve by herself. She often felt ill-equipped, especially when she felt people were trying to take advantage of her.

A few days before Thanksgiving, Raino Matson, who worked in the Stansbury Mine payroll office, came to our house to bring Daddy's final paycheck. Raino had been one of Daddy's closest friends. When he handed the check to Momma, she stared down at it and said, "Well, I guess this is it. I suppose I'd better start looking for another place to live." We all knew families had only a few days to move from the company-owned houses after they were no longer employed by UP Coal. He told her to sit tight until she heard from the company.

This was a very precarious time. Momma was in desperate need of money and was elated when a letter finally arrived from the insurance company, sure that it would be the double indemnity check. Instead, they requested a copy of Daddy's birth certificate because a newspaper article had reported Daddy's age wrong. The letter also stated that since Daddy was a miner, they would not pay double indemnity for an accidental death. Thus, Momma would have to wait even longer to be paid, and she'd get less. Once again we had to turn to our grandparents for money to tide us over. Finally, a letter from Equitable Insurance arrived containing a check for $5,000. Momma grasped the check in her hand and kissed it.

Then an incident involving a popular car dealer in Rock Springs got complicated. Six days before Daddy was killed, the car dealer had driven to our house in a new black and blue Oldsmobile hardtop that Daddy had admired on the car lot. I loved that car the minute I saw it. As Daddy and I came walking home from work one day, the dealer met us at the house with the Oldsmobile. Daddy was surprised to see him and said he couldn't afford a new car right then with the job situation the way it was, but the dealer persisted. "Just drive it around for six days, and if you don't like it, bring it back." Daddy let him drive our Hudson back to his dealership with the understanding that if Daddy didn't buy the new car we'd get the Hudson back. No one signed paperwork. Ironically, on the sixth day Daddy was killed.

The day after the accident, Momma called the dealer to arrange to return the Oldsmobile and pick up our Hudson. They set an appointment for her to come in on the day after the funeral. Johnny and I went with her that day, and heard the dealer bluntly tell her that they'd already sold our Hudson to someone else. The dealer explained that we would have to buy the Oldsmobile because it had been a brand new car and now it had miles on it.

Shocked, Momma blurted out, "But that's not what my husband agreed to, and you know it. So you'll have to get our Hudson back." The smug dealer, who wasn't so friendly anymore, waved off her concern and told her she could pay for the car with some of the insurance money she received. Everyone in town assumed Momma was going to get a lot of insurance, and many continued to believe until they died that she'd gotten a huge settlement. To pacify her, the dealer agreed to throw in free gasoline for one year, as long as we bought it at the Georgis Brothers filling station next to his dealership.

Momma was intimidated by this influential man—the biggest car dealer in town. The dealer wouldn't have tried this with Daddy, and if he had, Daddy would have seen to it that we got the Hudson back one way or another. But Momma had no paperwork and was in no shape for a fight.

I couldn't keep quiet any longer. I stood up. "You can't do this to my mother," I said indignantly. "It's wrong. You're nothing but a crook, and I'm going to tell everybody what you did." I stomped out of his office in tears.

I learned a good lesson that day; people aren't always as they seem, especially when money is involved. It would be a long time before I ever trusted a businessman again.

○ × ○ × ○

My mistrust was magnified the day I overheard Momma talking on the phone to our family dentist who had called regarding an unpaid bill for Daddy's dentures. Very emphatically Momma retorted, "I think I knew my husband, and my husband didn't wear dentures! How dare you try to get me to pay for something I do not owe." She slammed down the phone, and we never heard from that dentist again. It seemed like Momma had to fight for everything during this time.

One afternoon, out of the blue, Momma got a call from the administrative secretary at the UP Office in Rock Springs informing her that a company official wanted to meet with her early the next morning. "Well, this is it," she said after she hung up the phone. "They're going to tell me we have to move out of our house." She asked me to stay home from school the next day so I could go with her into town.

The next morning, after she got Johnny off to school, Momma dressed in one of her nice cotton dresses, and we drove into Rock Springs. She and I nervously sat in the waiting room. The door to the main office opened and Mr. Bayless, president of UP Coal from Omaha, came over to greet us.

Momma's mouth dropped when she realized she was meeting with the president himself. He put his arms around our shoulders and escorted us into his office, explaining he'd made a special trip from Omaha to see her. Our anxiety heightened as we took chairs in front of his beautiful mahogany desk. He leaned back in his chair and seemed to study us for a few moments before telling us how sorry he was about the accident. He said he'd admired Daddy and could only imagine how hard it was for Momma without him.

He paused and then told us the company was going to address some things. Momma gripped the arms of her chair, waiting for the bad news. He said, so that we wouldn't have to worry about housing, the company would let us stay in our house, rent free, for as long as we wanted. They would also deliver coal to our house each month, at no charge. Momma, who had prepared for the worse, was deeply touched. Tears overflowed her eyes and she said, "Oh, Mr. Bayless. You're a godsend. I was having an impossible time finding an affordable place for us to move into."

He told Momma that he hoped a rent-free home plus Social Security and Worker's Compensation would provide enough for us to get by. He got up and walked around his desk to where we were sitting and laid a hand on her shoulder as he handed her his business card. "If you ever need my further assistance in any way, please do not hesitate to get in touch with me immediately. Rest assured, I'll do whatever I can, but the one thing I wish I could do and cannot is to bring John back."

She went over to him and what started out as a handshake turned into a big hug as she thanked him profusely.

When we got outside the building, she stopped. "Can you believe it? God bless that man. This is a miracle, Marilyn. A miracle!"

"I know, Momma. I know," I replied. I too breathed a sigh of relief. "Now I'll be able to graduate from Reliance High School and have a chance at the honor scholarship. Mr. Bayless saved our lives today, Momma."

Later when Johnny came home from school, he jumped up and down at the news. "Oh, boy! Now I won't have to leave my friends." Then she called Jimmy and her parents. Grandpa, a retired UP miner himself, couldn't believe what the company had done as he'd never heard of UP being this generous before.

○ ✕ ○ ✕ ○

Right after school, Don drove over from Reliance in his parents' old car. As soon as he walked into our house, he asked me why I'd missed school. "I'll tell you all about it later," I replied.

Don ate supper with us that night, and after we did our homework we went for a drive and parked. He too was amazed at our good fortune. He pulled me over to the driver's seat, put his arms around me, and kissed me tenderly saying, "One of these days you won't have to live like this anymore." Then wrapped in one another's embrace, we talked and necked like young lovers do until it was time to go home.

○ ✕ ○ ✕ ○

The Safety Review Magazine *ran an article about our family and the benefits we received after Daddy was killed in the mine. The photographer encouraged us to smile but Johnny just wanted it over with. Jimmy was living in Rock Springs and was not available for the photo.*

Momma soon received a letter from Worker's Comp explaining her benefit was a flat $10,000, paid out in monthly payments until the youngest child was eighteen, whereupon the checks stopped. She wondered if we could live on Social Security and Worker's Compensation. It had been twenty years since Momma had been in the work force. She feared her typing and bookkeeping skills were rusty.

"Maybe I better call Emil Bertagnolli," Momma said. "The other day when I was in Union Mercantile, he said he could always put me to work. I told him I would have to see if Social Security would allow me to work and still receive benefits."

Momma started getting checks from Social Security. The Social Security Administration sent a reporter and a photographer out to our house to take pictures of our family and write an article that appeared in the monthly miners' *Safety Review Magazine.*

Momma agreed that facts about her Social Security benefits could be published in hope that it would help inform other widows. It stated, "Her husband has maximum earnings under Social Security; therefore, his survivors at his death became entitled to $203.10 a month. A lump-sum death payment of $255 was made to the widow in addition to these monthly payments: The three children, ages 10, 13 and 15 would each get $47.70 a month; Mrs. Nesbit would get $60.00 a month. [This is the breakdown of the $203.10, not additional sums, and was less than half of Daddy's wages.] Payments will continue to be made to Mrs. Nesbit and her children until the youngest child reaches age 18. These payments are to assist her in raising her children."

Momma's desire to help other mining families understand their benefits was stronger than her pride. Not many people today would allow their personal financial information to be made so public.

Much to her dismay, Momma learned that while receiving Social Security, she could only earn $1,200 a year. When she called Mr. Bertagnolli and explained, he said he could set up part-time hours around that figure. He also reminded her that by working even part-time, she'd be able to get a discount on things sold in the store: including groceries, clothing, shoes, and hardware.

○ ✕ ○ ✕ ○

One evening Jimmy got a ride from one of his friends and came out to visit us. After he'd told us all about school and life at Uncle John's he said, "I miss you guys, but I don't miss living out here. Look at this place." We all looked out the dining room window. "Look how desolate it is. The streets aren't even paved. It's ugly here. It's great Momma doesn't have to pay rent, but I wish the UP would close this camp down completely so Momma would be forced to move."

"Right now we are thankful just to have a roof over our heads." Momma explained. "Stansbury isn't that bad."

"Where are you going to get money to live?" Jimmy asked.

"Daddy's life insurance will pay off the funeral and the Oldsmobile, I'll cash in the bonds we had, and we'll use what Social Security and Workers' Compensation pays," she replied.

She then explained she planned to work part-time at Union Mercantile to make things a little easier for all of us. "Maybe you could stop by the store and see me once in a while, Jimmy," she encouraged. Then she added, "Right now I'm trying to think of things we can sell to bring in extra money."

o x o x o

Burt Dompson, a lady from the neighborhood, came over not long after that. She said, "Margaret, I've heard you're selling your piano and some things for extra cash and thinking about going to work at Union Mercantile, but why don't you try waiting tables at the restaurant where I work? The tips are great. Come with me tonight and give it a try for a few hours."

With some apprehension, Momma gave it a try but found it so overwhelming that it ruined her self-confidence and convinced her she might never be able to hold down a job anywhere.

o x o x o

Momma asked Johnny and me to go to the basement and get our silver aluminum Christmas tree and the ornaments out of the cubby hole. "We can at least try to make the house a little festive by putting up the tree," she said.

I found the box with the tree, but couldn't find the ornaments. "I can't find our boxes of ornaments, Momma. Did you put them somewhere else last year?"

"Did you look in that corner against the wall. They were always there."

"Well, they aren't there now," I replied.

A funny look slid over Momma's face and she said. "I think I know what happened to them. When Uncle Bob carried out the things I said he could have, he must have taken a lot of other things as well."

The next day Grandpa brought us some spare decorations. Don came over that evening to help put up the tree, and we set up the electric color-wheel that went with the tree. It rotated, casting alternating colored lights over the silver branches, which we thought was beautiful.

Afterwards Don and I drove around camp sharing our dreams. Don said over and over, "Someday I'm gonna take you out of this place," and the way he said it, I believed that he would. We talked about graduating from college, moving to a place like Salt Lake where the weather was always nice, and everything you planted grew. We'd buy a brick house like my Aunt Ann and Uncle Zeke's and live on a street set close up against the mountains.

The love Don and I shared grew deeper and deeper, and we ventured to express it in different ways. He was my first real love in all respects.

○ ✕ ○ ✕ ○

Right before Daddy was killed, he and Momma agreed to place a mail order through Spiegel's to redecorate the house. They'd decided to order drapes for the living room and sheer curtains, a bedspread, and chenille throw rugs for their bedroom and make monthly payments to the catalog company. Momma and I had spent hours looking through the catalog and folding over the pages of things we wanted. The day the order arrived, Momma and I were elated as we unpacked all the new items.

Now she could no longer afford the monthly payments to Spiegel's, so she asked me to help her box up everything so we could return it. I hated the thought of returning the items, but at the same time I didn't know what else we could do. We put the things in big boxes in the middle of the floor until Momma could call Spiegel's in the morning.

The next afternoon after school, Momma met me at the door and said, "Oh, Marilyn! You just won't believe what the lady from Spiegel's told me today. Spiegel's has this policy—if a breadwinner passes away, the bill is canceled. I just have to send her a copy of the death certificate."

"You mean we get to keep all the things we ordered?"

"Yes! Can you believe that?" So without any hesitation, we immediately unpacked the boxes and put all the things back in place.

I never forgot what that meant to us. From that day until this, I've always had a tender spot in my heart when it comes to Spiegel's.

○ ✕ ○ ✕ ○

One afternoon, right before Christmas, as the school bus pulled into town, I noticed a lot of cars coming into camp, and people rushing toward the mine. Then I heard the mine whistle blowing ominously. I searched among the kids getting off the bus for Johnny. He came running up to me. "Marilyn, what happened? Where's Momma?"

"Hurry! Let's get up to the mine." We ran straight for the mine, pushing our way through the crowd of people until we spotted Momma.

"Oh, my God, kids, someone's been hurt!" She wrapped us in her arms, pulling us close to her chest. "Don't look, kids! They're bringing up someone."

For a long time we just stood there, wrapped in Momma's embrace. Memories came swirling back as we wondered who it was that was injured this time and how badly.

The silence was finally broken when a friend of Daddy's came over to tell Momma that it was a trainer from Rock Springs. The trainer had reminded the miner he was training to always set the brake on the shuttle car. Then the trainer himself forgot to do just that, and the shuttle car flew away down the slope. The man was injured, but easily could have been killed.

o x o x o

I was awakened sometime after midnight to the sound of Momma crying in her bedroom. Quietly I got out of bed and went to see what was wrong. She was just sitting on the edge of her bed looking at the picture of Daddy on the dresser and clutching in her hand his gold watch and black onyx ring.

I startled her when I walked into the room. "I didn't mean to wake you, Marilyn. Sorry. Hearing that whistle blow again and being at the mine today rekindled all the memories. I just felt like holding onto something of your father's for a little while. Whenever he got all dressed up, he loved wearing this watch and ring. God, how I miss him," she said as her head slumped down and tears streamed down her cheeks.

For a while we both sat at the foot of the bed in silence, each with our own thoughts, and then she turned to me and said, "Go back to bed, Marilyn. I'll be all right."

o x o x o

Then one night, long after we had all gone to bed, I woke up and noticed a light dimly streaming under my bedroom door. Curious, I quietly put on my robe and tiptoed to the kitchen where Momma sat huddled over the kitchen table, writing a letter.

"Momma, what are you doing writing a letter in the middle of the night?" I asked.

"Oh, Marilyn! You startled me," she replied. "I couldn't sleep for wondering what we are going to do."

"What do you mean? I thought we were making it all right," I said.

"There's something I've been keeping from you kids. I've heard that the UP is going to officially close Stansbury. I'm not sure exactly when, but I have to start doing something now because they're planning to sell all the houses in camp. If they do that, we won't have a house. Having this house rent-free has been the only way we've been able to make it financially.

"So I've been making calls, trying to find us a place we can afford to rent in Rock Springs, but there isn't much. When I do find something, it's a tarpaper shack or the rent is too high or they won't take a single mother with three teenaged kids."

"Well, we can go live with Grandma and Grandpa if we had to for a little while, can't we?" I asked.

"Oh, Marilyn, they don't have room for all of us. That's why I have to write this letter to Mr. Bayless in Omaha. Remember, he told me to get in touch with him if ever I needed anything. But I don't know what he can do for us since the company is going to sell all the houses. With the houses gone, there won't be a Stansbury anymore. I know this is probably a waste of time, but for some strange reason I just felt compelled to write this letter to him tonight."

"But Momma," I questioned, "What can he do? We don't have any money to buy the house we are living in."

"I just have to try," she replied. "Now I want you to go back to bed and pray. Do like Grandpa always does, fall to your knees. Pray really hard, Marilyn, that our prayers will be answered."

I went back to my bedroom, took out my diary and wrote about what Momma was doing this very moment, and underlined it in red. Then I knelt down beside my bed to pray.

Long after Momma had finished her letter and gone to bed, I quietly got up, went to the phone, and dialed Don's number. When he answered, I told him that the company planned to sell all the houses in Stansbury. He told me not to worry; it was probably just gossip.

CHAPTER TWENTY-THREE

FATE STEPS IN

MOMMA MAILED her letter to Mr. Bayless. Every day for two weeks after that, she went to the post office, hoping to receive a letter with a Union Pacific return address but that letter never came.

One afternoon, when she had just about given up, she received a phone call from the UP office. Mr. Bayless was in Rock Springs and would send a car out to pick up Momma. Again she wanted me to go with her, so I didn't go to school that day. Within a short time, Vern Murray, a UP official who had been a friend of Daddy's, arrived to pick us up. After we got in the car he turned to Momma and asked in an angry tone, "Why are you always asking Mr. Bayless for so many things?"

Shocked, I stared at him in disbelief. Momma looked at him and replied in a hard tone that left no doubt how she felt. "Vern, I can't believe I heard you say that! My business with Mr. Bayless is none of your concern." Not another word was said to Mr. Murray from that moment on. Here he was, a highly paid UP official who wanted for nothing. He was someone Daddy respected, and who both my parents knew well, yet he seemed to be resentful that UP was helping us.

○ × ○ × ○

Mr. Bayless was a handsome, older man, polished in every way. He always wore a tailored suit with a vest and a tie. Anyone in his presence knew he was "the boss."

As Momma and I entered his office, I remember comparing us to the prince and the pauper, we being the paupers. I wondered why this important man had sent a company limousine to pick us up, when he could have responded to Momma's letter via mail. I knew he had to be busy and really didn't have time to be concerned about the welfare of one miner's family, but I was wrong. He did care.

He started the conversation by telling Momma he had received her letter and wanted to respond to her in person. Momma and I sat quietly waiting to hear what he had to say. With a disheartened look on his face, he confirmed the fate of Stansbury. He explained that over the next couple of years the mine would be closed for good, all the houses would be put up for sale to the public, and the new owners would be required to move them from Stansbury. Momma sat motionless and said nothing. Finally I said, "What's going to happen to us?"

"Marilyn," Momma hushed me, "let Mr. Bayless continue." But before he could say another word, Momma asked, "Then you're telling me I have two years before I have to move out of our house?"

"That is what I want to talk to you about today," he replied. I waited for the hammer to fall.

He asked that everything we talked about be held in confidence, explaining that the public would hear about the closure through an official announcement later. "I once promised you that you would always have a place to live, and I intend to keep that promise," he assured us. He then explained that the company had made arrangements for Momma to own one of the company houses.

"What do you mean, own a company house?" Momma asked in bewilderment.

"You'll be able to own one of the houses," he replied.

"I need to ask you a couple questions," he went on. "Before any houses go up for sale, I want to know which house you want. I think it would be best if you continue to live in your current house for now and choose another house to be moved. That way the second house can be moved, set up, and waiting for you in Rock Springs when it is time to move from Stansbury. We have been looking at a couple of potential developments in Rock Springs—one in the Number Four district and the other in what is to become the Bellview Addition."

"The Bellview Addition? Where will that be?" Momma asked.

"It'll be on the hill within the city limits just north of Rock Springs. The investors who plan to purchase the majority of the Stansbury houses

I.N. Bayless served Union Pacific Coal Company in several positions in Wyoming, becoming president of the company at headquarters in Omaha. (Sweetwater County Museum)

will develop it. Other buyers of Stansbury houses will have the option of moving them to the Bellview development as well."

"I grew up in the Number Four district so I'd prefer the newer Bellview area," Momma said.

"Fine. Once your house is moved, it will be set on a foundation, a furnace installed, water and utilities hookups brought in, so all you have to do is move in. Understand, we are on the ground floor with this, so I don't have definite start-up dates."

Mother took a handkerchief out of her purse and dabbed her eyes, trying hard to hold back a swell of tears. She lowered her head and thanked Mr. Bayless again and again. A heavy burden had been lifted off her shoulders, and I felt Mr. Bayless understood that as he sat waiting.

Finally, Momma couldn't sit quiet any longer. "That night I wrote you, I didn't know what to expect but something made me write. I didn't know where to turn. I knew if anyone could help me, it would be you. I don't know what we would have done without you standing by us through all this," she said. "Again my prayers were answered."

Mr. Bayless looked across his desk at me and said, "Marilyn, over the years, I remember seeing you with your father when I was in Stansbury. First you were just a little girl, and now you're growing all too fast into a lovely young lady. Come here beside me for a minute." I walked to where he was sitting, whereupon he stood up and put an arm around me. "Your father was always so proud of you and your brothers. Your mother needs all of you right now. One of these days all this tragedy will be behind you, and you kids will go on to become successful individuals. I want you to know how horrible I felt the day your father was killed, and I'm pleased to play a small role in helping you rebuild your lives. Now, young lady, which house do you think your mother should choose?"

"Oh, Mr. Bayless, I love the house we're in because my bedroom is painted and wallpapered just like I wanted.

"Well, as I said, we can move the house you are currently living in if you want, but it seems like it would be easier if you chose another."

"Whatever Momma wants to do is fine with me. I won't be needing that bedroom much longer anyway. I'm just feeling sad about Stansbury. Where will the people go? I was worried about us, but since the company is giving us a house, we'll be fine. But what about everyone else?"

He gave me a tight hug and said, "I hate to see this mine close, too. But my hands are tied. There just isn't a big demand for coal anymore." Then, to indicate it was time for us to go, he said, "Now, Mr. Murray will take you back home."

I blurted out, "Could you have someone else drive us home? Mr. Murray was rude."

"Oh? What did he do?" Mr. Bayless asked.

"He said my mother is always asking you for too much." I replied.

Mr. Bayless quickly retorted, "I'll have a little talk with Mr. Murray, but you will have to ride with him one last time. However, I assure you he will be polite. Please excuse me for a moment."

○ × ○ × ○

When he returned, he walked Momma and me to the door. He asked, "Now tell me, Marilyn, did you like growing up in the camps?"

"Oh, yes! And I am so glad we get to keep living there so that I can graduate from Reliance High School and get an honor scholarship."

"Honor scholarship?" he asked.

"Yes. I might not have gotten one of those scholarships at another high school, and then I wouldn't be able to afford college. I've always wanted to go to college. I know I'll have to work my way through, but if I can get that honor scholarship, at least I won't have to pay tuition."

Mr. Bayless walked us to the waiting limousine where Mr. Murray sat behind the wheel in quiet obedience. Mr. Bayless asked Momma to call him when she decided which camp house she wanted.

○ × ○ × ○

I never saw Mr. Bayless again, though I would remember for the rest of my life what he had done for my family. Also on that day I realized what a courageous person Momma was. Up until then, we'd looked to Daddy for guidance, wisdom, and strength. But now Momma had stepped into this role, something she never wanted or dreamed she'd have to do.

Momma chose the house the Besso family had lived in, to become our new home, and we would soon begin sorting through eleven years of accumulation to prepare for the move.

Gossip ran rampant regarding the fate of Stansbury though no official announcement had been made. Neither Momma nor I breathed a word of the information Mr. Bayless had given us, but suddenly rumors began circulating that Momma would buy the house we lived in if Stansbury closed. People were convinced she had received a big insurance settlement that paid double indemnity for accidental death. "How else could she afford the house?" they theorized. They all knew for certain that UP had never treated anyone kindly—and never would. So it had to be that Momma had used some of her huge insurance settlement to buy the house. Many seemed almost envious that we would have some-place to live, while they did not.

○ ✕ ○ ✕ ○

As an attractive, forty-year-old widow who was assumed to have money, Momma faced new problems. She became a romantic target for both single and some married men. Married men who had been Daddy's friends began coming to our house under the pretense of helping her fix things. She often bore the brunt of scurrilous gossip and false accusations simply because of her status.

Momma started work at Union Mercantile. She loved everything about this job, and the extra money really helped out. We did all our shopping there and, even though customers could run charge accounts, Momma paid cash for her purchases. Occasionally we still experienced those dreadful times in the grocery line when I had to take groceries back because we didn't have enough money. I became accustomed to ignoring the looks of people standing in line behind us when this happened.

We kids got used to Momma's working and not being at home waiting for us after school. Sometimes, when the weather was bad, she'd call and ask, "Kids, the road is too icy for me to drive home alone tonight, so I'll have to stay at Grandma's. Marilyn, can you and Johnny manage on your own tonight?"

And what could I say? "Oh, sure, Momma, we'll be fine." I'd keep the furnace going and fix supper for me and Johnny.

On these nights, Don drove over from Reliance and the three of us sat around the kitchen table doing homework, or we'd help Johnny memorize his spelling words. Then, we'd pile into Don's car and ride around the snowy roads in camp. When it was time for Don to go home, Johnny and I hated to go back to our empty house. We felt so alone, but we never admitted our fears to Momma. We tried to be brave for each other but we left the living room lamp on all night.

One night when Momma wasn't home, Johnny and I started a nightly ritual, which we carried on until the day we left Stansbury. Due to the mine layoffs, the camp was almost abandoned. Houses stood empty and streetlights had burned out. At night, the whole camp was dark with shadows where they had never been before. Strangers drove through camp with their headlights turned off, looking for things to steal from the boarded-up houses.

Johnny and I turned off our house lights, called Tippy inside, and watched out my bedroom window to see what was going on in camp in the night. We saw men get out of dark vehicles and load things into the backs of their pickups. They took just about anything: old garbage cans, pickets and gates from fences, clothesline poles and wires, screen doors, front and back doors. Some even dug up trees and loaded soil from yards and gardens. We could hardly breathe if a truck pulled up to an empty house close to ours. Each night we took butcher knives from the kitchen and slept with them underneath our pillows. I'd lie in bed with my hand gripping the knife handle and my heart throbbing, listening for scavengers to open the door into our house. I would finally get so tired that I fell asleep. Johnny and I started sleeping with the knives under our pillows even when Momma was at home.

Momma worried about our being alone in the camp after school until she got home from work. One day Johnny and I had just gotten home from school when a Union Mercantile delivery truck drove into our driveway. The driver told us that he was delivering a television set. We were sure that was a big mistake so we called Momma at work to see if we should let him in the house or not.

Words cannot begin to describe how excited we were when she said, "There is no mistake, kids. I thought we'd gone long enough without a TV, so I asked the manager if I could make monthly payments. Tell the delivery man to hook it up under the window by the front door. And be sure he brought the TV light. Be sure to turn on that light or it will ruin your eyesight." People believed at the time that watching television in the dark would damage your eyes, so they placed small figurine lamps, similar to night lights, on top of the TV. Momma also had purchased a three-colored sheet of cellophane—blue across the top, red in the middle, and green across the bottom. We Scotch-taped it over the screen of the black and white TV and loved the way it made the picture look colored. That television was the greatest thing Momma could ever have given us. No longer did we dread going home to the empty house after school each day.

○ ✕ ○ ✕ ○

One afternoon my friend Geraldine said, "I've got some news. My family is going to move to Walsenburg, Colorado, by the end of the month. My dad has a job in a coal mine there." My heart sank. Geraldine had been my friend since grade school. First, my friend Carolyn had moved and now Geraldine. I was devastated when their moving van pulled away.

I missed Geraldine so badly. I found myself walking by their empty house every day when I got home from school. I wondered what Walsenburg was like; was it prettier than Stansbury and was there more to do there? Then the UP men came and boarded up Geraldine's house. That night, I went down to the abandoned schoolyard and sat in one of the remaining swings on the playground, all alone, until the sun went down. As I pumped the swing higher and higher, I thought of all the time she and I had spent here playing on the swingset. Wonderful memories filled my mind, while tears streamed down my cheeks. The camp was so quiet now. Everywhere I looked, houses were boarded up. The street light in the schoolyard had burned out, and soon it turned pitch black all around me. I jumped from the swing, high in the air, and hit the ground running. I ran back to our house afraid to look back for fear a stranger was following me in the dark shadows.

My life wasn't the same without Geraldine. When I wasn't with Don, I spent most of my time in my bedroom playing the 45 RPM records Geraldine and I had once sung along with. The ache I felt was almost physical, and it was painful.

I wrote more and more letters to Aunt Ann and Uncle Zeke, and wrote things I never could have said out loud. I wrote that I didn't know what was going to become of us and wondered why God let this happen to our family. I shared with Aunt Ann my innermost feelings about how sometimes I wanted to get away from it all and live with them, but then I realized this was not an option for me. What was left of our family had to stick together, no matter what happened. She wrote back or called as soon as she got my letters, but she never again pressured me to leave, and for that I was grateful.

○ ✕ ○ ✕ ○

Jimmy, Grandma Copyak, and Johnny standing by Daddy's headstone on the first Memorial Day after he died.

Then came the unthinkable: Momma started dating. No one could ever take Daddy's place for Johnny and me, and we expected her to feel the same way. I knew how lonely she was, but I was devastated. I challenged her and she replied, "Marilyn, I've got my life to live, too. It isn't easy walking through life alone."

We did not like Momma seeing anyone else. "If you want to date other men, wait until I graduate and leave home." I sobbed, selfishly. It reached the point where Don said he felt sorry for Momma and that I shouldn't carry on like I was. But I was genuinely worried about her when she went out to dinner or to a movie and wondered what we would do if someone harmed her. On my urging, Don agreed to follow her wherever she went on a date. When Momma found out what we were doing, she was livid!

○ x ○ x ○

The day before the first Memorial Day after Daddy's death, Momma, Johnny, and I went into Rock Springs, picked up Jimmy at Uncle John's, and went up to the cemetery to spruce up Daddy's gravesite. We hauled water from home to water the lilac bushes, wiped down the headstone, and picked up debris. The cemetery was full of people and many stopped to share their memories of Daddy.

Before we left we visited the dollhouse monument. The legend was that this grave marker was built in the 1920s by a father who had lost a child. About four by five feet, built of crushed stone and mortar, it resembled a playhouse or dollhouse. The roof of the small, open porch was supported by concrete pillars. Two urns stood near the door. Glass covered the window and curtains hung inside it. Someone had put a doll and other toys on the inside as if ready for little hands. We children speculated endlessly about who had built the monument and who was buried there.

As we left the cemetery, we were very quiet as we pondered a world in which not only are adults taken from families too soon, but also where parents must sometimes must bury their children.

Rock Springs historical researchers have recently discovered that two of the ten children of Henry C. and Linda Livingston rest there. Henry Livingston had lovingly hand-built the dollhouse to mark the final resting place of the children, Henry Wayne and Georgia Rozena Livingston, who died of diphtheria or perhaps spinal meningitis.

○ ✕ ○ ✕ ○

Early Memorial Day morning we went back to the cemetery. Even at that early hour, the cemetery was again full of people. We placed a red carnation arrangement in front of the family headstone and one red rose across Daddy's marker. On previous Memorial Days, Daddy had joked, "When I die, I don't want anyone wasting money on flowers for me. All I need is one red rose."

CHAPTER TWENTY-FOUR
A WAKE FOR STANSBURY

F OR ALMOST ninety years Sweetwater County's black gold had fired coal-burning locomotives. When UP began shutting down mining operations in the mid-1950s, the company at first kept the Stansbury and Superior mines in operation on a standby basis. Stansbury had been idle for several months in 1956, but reopened during July of that year when an increase in rail traffic resulted in a number of coal-burning locomotives being returned to use.

But by February 1957, time had run out for the remaining two hundred Stansbury miners. On February 20, thirty-six Stansbury employees received notices that they were transferred to the Superior mine. No other information was given to them regarding the transfer. The thirty-six men were those with the most seniority—mostly mobile unit operators, a couple of repairmen, a hoist man, and two or three motormen. Some supervisory personnel were also transferred. When word of this got out, Stansbury miners and their families were anxious and worried. UP Coal President Bayless confirmed from company headquarters in Omaha that transfers had been made but denied rumors of layoffs at Stansbury. V.O. Murray, vice-president of operations, the same man who had been rude to my mother as a driver, declined to confirm or deny widespread rumors that the mine would be closed. Bayless said from UP in Omaha, "We still don't know yet what we are going to do."

On February 22, only two days later, as the thirty-six men were being transferred, a statement from Mr. Bayless was posted on the Stansbury bathhouse door: "The Stansbury mine north of Rock Springs has been closed indefinitely as of February 22, 1957. Present conditions for bituminous coal supply do not warrant continued operation of Stansbury. It is an action we take with regret, but present conditions force us to do so." I remember that day as if it were yesterday. This notice hit hard. Miners and their families all knew it was eventually going to happen, but they felt that UP was devious about their plans. In practically every house still inhabited in Stansbury, people cried. Momma, of course, had known it was coming since the day Mr. Bayless asked her to pick a house to live in. She had kept his confidence.

This was a devastating blow to the miners, who lived paycheck to paycheck, armed only with their well-honed mining skills and little formal education. The heartache I witnessed in the homes of these families was beyond words. Hardworking men sat huddled with their families around kitchen tables, crying, and praying that they would be part of the skeletal crew needed to work during the final closure of the mine.

This company town had instilled in its residents a sense of belonging and a common focus. Now, the bond of town and mine was gone. The remaining families were now forced to do the unthinkable: apply at the Wyoming Employment Security Commission Office for jobless benefits beginning the week ending February 23.

Previous to this, when a family moved, UP had immediately boarded-up the vacant house making it stand out as a target for nighttime vandals. What was it going to be like when all the houses were boarded up? Would it be safe to live in Stansbury then? Women and children were afraid to venture outdoors after dark.

The dissolution of Stansbury continued when Louise Tomassi, camp UP store manager, put this advertisement in the *Rock Springs Rocket-Miner*: "Notice: I am closing the Stansbury store. All merchandise in stock will be sold to the general public at wholesale prices. Closeout sale will start Monday, March 11 and continue to April 1. Drive out to Stansbury and save! All sales final, no refunds, no exchanges."

At the end of March, the remaining Stansbury residents and many former residents living in nearby communities gathered together on a Saturday night in the camp's community hall for a potluck supper, a community sing, and a dance. They reminisced about the good times and the sad times. It seemed ironic to outsiders that in this time of loss, the community reacted by having a party, but to us it was a final fling, a goodbye party, a "wake" for a community and lifestyle that had met its end.

After that, except for us, the families began leaving Stansbury for towns like Rhanda, West Virginia; Walsenburg, Colorado; Chicago, Illinois; Long Beach, California; Price and Vernal, Utah; Carlsbad, New Mexico; Kelso, Washington; and nearby places in Wyoming. Some found mining jobs, some moved back with family, others took jobs doing anything to support their families. As one miner said, "I'll pump gas to feed my family if I have to."

I don't know how Momma managed to keep her spirits up, knowing that soon the only people living in Stansbury would be us and the company watchman. At times she cried, but she never faltered. Throughout this time, she had the uncanny ability to make my brothers and me feel special and well cared for.

Though Union Pacific closed the Stansbury mine, it maintained ownership of the vast lands, the untapped wealth below, and the railroad spur. Thus, in the minds of corporate management, this area must have still had some value; it had the potential to become marketable or profitable again with the passing of years.

○ x ○ x ○

I worried that Reliance High School would not survive long enough for me to graduate there. I wondered if I would be able to hold my own in the much larger high school in Rock Springs where more students would be competing for the few scholarships.

On May 3, 1957, Don and I signed up to decorate the high school gym for the junior prom. Our class chose "The Stork Club" as the prom theme, complete with the skyline of New York City decorating the walls of the gym. It truly would be an escape from reality.

Momma paid for me to have my hair styled at the Rialto Beauty Shop owned by Onnie Pastor. Kay Grillos, my favorite hairdresser, fixed my hair beautifully. The night of the prom, I wore the only "formal" Momma and Daddy ever bought me, a mint green dress with a white net skirt overlay. I felt like a queen.

Don drove over at six that evening in his parents' new Oldsmobile. He handed me a floral box containing a beautiful orchid corsage, and

I was selected queen of the Junior Prom at Reliance High School. Shown: Joan Nichols, Marilyn Nesbit, Patsy Daniels, Leona Theinpoint, and little Patty Pirnar. (Reliance High School Yearbook photo)

Momma scurried to pin it onto the left side of my dress. I looked into the eyes of this strikingly handsome man, pinned his baby's breath boutonniere on the lapel of his new suit, and knew I was blessed to have him in my life.

In spite of declining enrollment, students, teachers, and chaperones filled the gym and danced to the music of a live band. Don grabbed my dance card out of my hand and scribbled his name on every line. At intermission, the superintendent announced a senior, Donnie Keelan, had been voted prom king, and my heart danced when I heard my name called as queen, even though I was disappointed that my Don wasn't chosen as king.

Later Don and I parked in our special spot and looked over what was left of Stansbury. I could pick out Momma's porch light glowing from the darkened town. A bittersweet feeling came over me, and I thought of the beginning of the book *The Tale of Two Cities* that we'd read for English: *"It was the best of times, it was the worst of times. . . it was the season of Darkness, it was the spring of hope, it was the winter of despair, we had everything before us, we had nothing before us. . . ."* Tears flowed down my cheeks.

○ ✕ ○ ✕ ○

Days turned into weeks and weeks into months. And before I knew it, I was in my senior year of high school. Don and I continued to go everywhere together and see each other just like always. I could not distance myself from the thought of a life with Don and me together forever. I loved him more than anything in the world, but our relationship was now carried by the wind.

About three-quarters of my freshman class had moved away by the time we entered our senior year, including my two best girlfriends, Carolyn Pecolar (top) and Geraldine Guigli (bottom).

CHAPTER TWENTY-FIVE
NO REGRETS

THE DAY THE photographer took our group class picture at the end of my senior year was very somber. My senior class had dwindled to only twelve students from forty-one students when we were freshmen. Rumors had it that we would be the final graduating class.

At the final awards assembly, Superintendent Chadey announced the recipients of the awards: athletics, drama, music, photography, math, and the honor scholarships. I held my breath in anticipation. Carl Pearson was awarded the valedictorian scholarship, and I was awarded the salutatorian scholarship. Don and I were determined to enroll at the University of Wyoming in the fall, but no one else from our graduating class planned to go to college.

Aunt Ann and Uncle Zeke arrived a few days early to attend my graduation ceremonies and stayed at Grandma's house. Excitedly I telephoned them.

Aunt Ann asked, "So, you're still planning on going to the University of Wyoming?"

"Yes, I can hardly wait to go to Laramie at the end of the summer. I'm working again at Jay's Drive Inn this summer. I saved most of the money I made last summer, and I've been offered a part-time job working for the head of the business college at UW. It's all working out," I replied.

○ x ○ x ○

My Latin teacher, Regina Black, was helping me with my Salutatorian Address. I wanted to pay special tribute to Momma and Daddy for all they had sacrificed to get me to this point. Miss Black thought I was being too sentimental and steered me in another direction by saying

"Sometimes, to see the future, we have to turn our back on the past." So, heeding the advice of this woman I admired, I wrote about the space era, rocket ships to the moon, and the future of our society. That was a mistake. I felt far removed from the space age. My thoughts were on the coal camps and the families who had resided in them, and the part they had played in my life. With everything that was going on around me, is there any wonder I didn't feel connected to the space age?

On the night of graduation I took the podium, I glanced down into the crowd to make eye contact with my family members. Daddy's absence hit me immediately. I could almost see him in his best suit, sitting by Momma and looking pleased.

Standing at the back of the room was Dr. Muir. Tears welled in my eyes when I saw him. Then I remembered Momma cautioning me, "Whatever you do, don't cry during your speech. Be strong!"

I delivered the space age speech, but before I sat down my inner feeling bubbled to the top and I ad-libbed about how proud I was to be a coal miner's daughter. I could feel Daddy's presence and knew how proud he would have been. Then I stepped back to join my class and took a seat next to Don. I dropped my right hand to my side and felt his comforting hand entwine mine.

After the ceremony, my classmates and I lined up along the gymnasium wall to receive congratulations. My arms soon became laden with graduation gifts, and I had to pile them on the floor behind me. Don's mother and father thanked me for being a good influence on their son.

When everything was over, Don left with his family, and I went to my grandparents' house with my family. After dinner, Aunt Ann took me aside and said, "We were so proud of you today. Your speech brought tears to my eyes. I know you won the honor scholarship to the University of Wyoming, but I'm curious as to whether it will be enough?" I explained to her that it paid $125 per semester, enough for tuition and fees only.

"Well, your uncle and I have an idea for you to think about. We are moving to Denver this summer. We'd like for you to come live with us and attend Denver University. It's a more prestigious college, you know, and would look impressive on your resumé when you go to get a job. It's an opportunity not many girls get."

"Aunt Ann! Thank you, but I've already decided. Don and I want to go to the same school. We've been planning for years, so please don't try to change my mind. "

When we got home later that evening, Don's car was parked in our driveway. He gave me a quick hug and I said, "Come inside. Give me a couple of minutes to change my clothes. Then we'll go for a ride. I have so much to tell you!" As Don and I drove up the winding road leading to our special place, I told him of Aunt Ann's offer.

Don parked the car and turned toward me, "You wouldn't leave me to go to Denver, would you?" He looked like he was about to cry.

"Oh, Don, no. I never gave it a thought!" He pulled me over to him and held me close, burying his head on my shoulder.

We didn't say a word for a while, and then he said, "You had me worried for a minute. I wish we could get married right now. I'm working for the power company again this summer, and I'll save my money. If you saved the money from your job, we'd have enough to get married. It would be cheaper if we got married, because then we'd only need one apartment rather than two dorm rooms."

"Oh, Don," I replied, "the four years will go by in no time, and then we can get married. We'll have the big wedding I've always dreamed of having. Four years, that's all it is! Then we'll both be out of here. Look down there. Look how drab Stansbury looks, houses boarded up, street lights broken out, no paved streets. I don't ever want to live in a place like this again, where the company can just come in and close down the whole town. We can get good jobs and buy a nice house in the city. Some place where the wind never blows. But we have to wait to do everything just right. Things will fall in place for us!"

For a while we quietly kissed and snuggled until we were lost in each other's embrace.

○ × ○ × ○

And so my last summer in Stansbury began. On the weekends when I wasn't working, I helped Momma pack. Even though we did a lot of packing, I knew there would be so much more to do before Momma made the final move. I just hoped my uncles would help her because there was no way she, Jimmy, and Johnny could do it by themselves.

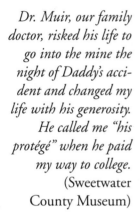

Dr. Muir, our family doctor, risked his life to go into the mine the night of Daddy's accident and changed my life with his generosity. He called me "his protégé" when he paid my way to college. (Sweetwater County Museum)

I knew Momma couldn't help me financially when I went to college. My fears heightened when I learned that the money she received for me from Social Security and Worker's Compensation, $46 and $25 respectfully, had stopped the previous January, when I'd turned eighteen, five months before I graduated.

○ × ○ × ○

I spent the summer working at Jay's Drive Inn and loved the work. I saved as much money as I could. The boss, Pete Skitch, was a good man and asked only three things of his employees: come to work on time, always wear a hairnet, and eat any mistakes you make. That sounded great, until I made a couple of big mistakes and had to eat the orders I'd screwed up.

Don and I continued to see each other every evening, always ending up at our parking place on the hill. We couldn't have spent more time together if he'd lived at my house. We talked about everything, and with his help I began sorting through my feelings about the accident.

At the end of the summer, I quit my job to get ready to leave for college. Aunt Ann had continued to send me all kinds of clothes. Even though I was the poorest girl around, I had more clothes than anyone I knew. The last thing on my college to-do list was to go to the Miners' Clinic and get the physical required by the university.

The clinic was filled with patients but within a few minutes, Dr. Muir's pretty, red-headed nurse, Rusty, came into the waiting room and called, "Marilyn, Dr. Muir is ready to see you now."

Dr. Muir was forty years old, a tall man with kind eyes and the manner of a southern gentleman, 'though he'd never lived in the south.

After my examination, he inquired as to how I had been, how my family was, how I liked my summer job, and complimented me on receiving the Salutatorian Award at graduation.

"Thank you for coming to my graduation. I saw you standing in the back of the hall, but you left before I got a chance to talk to you," I said.

"I was called back to the hospital. What do you plan on majoring in at the University?" he inquired.

I told him business education and then he asked, "How are you financing your education?"

I answered quietly, "Dr. Muir, receiving that scholarship made it possible for me to even consider going on to college. After Daddy was killed, I thought college was out of the question. I'm not sure I'll be able to make it, but I've got to try. The scholarship pays for tuition and fees. I've saved $600 and Momma gave me $200 she saved from the money she got for me from Social Security and Workers' Compensation. I have a job lined up, working for the Dean of the College of Business. So I should be okay."

"Well, it sounds like you have everything planned out. You know working and going to college at the same time is tough, but somehow I think you'll make it." He paused a moment and then asked, "Could you and your mother stop by my office again tomorrow?"

Taken aback, I said, "I think we can, but I'll have to ask Momma if she can arrange her work. But why do you want to see us both?"

"There's just something I want to discuss with you and your mother."
I couldn't imagine why he wanted to talk to us. Did I have something
wrong with me that he didn't want to tell me without Momma being
there for support?

Standing up from behind his desk, he walked around his desk. "If
your mother can't come, give me a call and we'll work something out."
I thanked him and left. Dr. Muir had seen us kids through all of our
childhood illnesses, had performed surgery on both my parents, and I
could never forget that he'd risked his life the night of Daddy's accident.

○ ✕ ○ ✕ ○

It rained really hard the next day when I had so many last minute things
to do. I called Momma at work and asked, "Do we have to go see Dr.
Muir today? Can't you just call him to see what this is about?" I was a
little apprehensive about this mysterious appointment.

"No, Marilyn, this must be important. Get ready, and I'll be out to
pick you up in a few minutes," Momma instructed.

When we walked into the clinic, it was busier than ever. Every chair
in the waiting room was taken. It was so crowded that Momma and I
joined other patients leaning against the wall, waiting to be called.

We waited and waited. Finally Rusty appeared in her crisp white
uniform that rustled like dry leaves when she walked, called out my
name, and motioned for Momma and me to follow her. As we walked
alongside her, she chatted with Momma and then turned to me and
asked, "Well, how are you today, Marilyn? Probably excited about school,
huh? Dr. Muir is so anxious to see you."

"Rusty, do you know why he wants to see me again," I asked.

"No, I don't," she replied softly.

Rusty tapped lightly at Dr. Muir's door, and he greeted us cordially
as he led Momma and me into his office. He motioned for us to take a
seat in one of the upholstered chairs in front of his desk. When he walked
past me, he put his hand on my shoulder before he walked around his
desk and sat down. "For a while there, I was beginning to think I might
have to reschedule your appointment. We've had a waiting room full of
people since early this morning," he said in his quiet, gentle voice that
had consoled me so many times before. "Sorry you had to wait so long."

He leaned back in his chair and seemed to be studying Momma and me for a while. I felt my throat closing up. Finally he began "Marilyn, I've been following you closely since your father's accident. Saw in the paper you were chosen prom queen, participated in music festivals and debate trips, won a scholarship, and even gave the Salutatorian Address at your graduation. I know your father would have been very proud of you. I just wish he were here to share your life with you. I'm curious. Did you ever get a chance to tell him you wanted to go to college?" he asked.

I lowered my head as I explained, "Daddy thought a girl didn't need to go to college. But when I kept asking him about it, he told me the only way I could go on to college was if I earned an honor scholarship. When he got killed, I promised myself I'd get that scholarship no matter how hard I had to study. And I got it. "

Dr. Muir nodded, "You know, it's always been my feeling that there are times in everyone's life when they can use a helping hand. So, if your mother will allow me, I would like to offer you that helping hand. It would mean a great deal to me if I could do that." The room went quiet. I turned to Momma in bewilderment, waiting to see her reaction because I had no idea what he meant.

"Since your father isn't here, I would be honored if you would allow me to fill in for him as far as your education is concerned." He reached into his top desk drawer and took out a leather checkbook and slowly slid it across the desk toward me. My parents paid cash for everything, so I'd only seen checkbooks when other people wrote checks. Suddenly I realized he was giving this one to me.

"I have taken the liberty of setting up a personal checking account in your name. Since I don't know exactly what my own day-to-day expenses are, I cannot expect you to know yours. Therefore, I put enough money in this account to take care of your everyday needs. Spend it any way you like. I never want you to be deprived of anything important just because of money.

"But I ask that you do three things for me. I don't want you to get a job while you're going to college. I want you to enjoy college life so I'd like you to join a sorority and take part in all the activities they offer.

Also I don't think you should get involved with just one boy. Go out and meet new people before you get serious about anyone." He looked at me in a very serious manner. "Do you think you could do these things for me?"

My eyes must have widened as I thought about Don. "Dr. Muir, I've gone steady with the same boy all through high school."

"I realize that," he replied. "But that was high school. Things change when you start college. I'm not asking you not to see Don. What I am asking is that you see other people as well. I don't want you to be reluctant to get involved in sorority functions because you're going steady with someone. Go out! Have fun! Find another life after Stansbury." He stood up. "Now, do either of you have any questions?" he asked.

The room again went quiet. For a while none of us said a word. Spellbound, I asked, "Do you really mean what you've just said? Do you know how much money that would be, especially if I don't work?"

"Yes, I think I know how much money that could be," he replied with a smile on his face.

"Why would you do this?" I asked.

"Let's just say I have my reasons," he replied. "You've always been a special girl, even when you were little, with special parents. You have a wonderful, courageous mother. Your dad! Well, your dad was a very rare individual. He worked so hard and went through a lot for his family, and now he isn't here to open doors for your future but I can help. I want you to walk through those doors, with your head held high," he said enthusiastically. Then, as if wondering if he'd said too much, he hesitated and said, "I just hope I haven't offended either of you by making this offer."

"Offended! Oh, Dr. Muir! I was so worried about money, but I didn't want anyone to know. I can't believe it. How could I ever repay you?" I asked.

"Just repay me by being a good person, compassionate, and always, always conduct yourself like a lady wherever you go," he answered. When he uttered those words, he sounded just like Daddy. In a few short moments, Dr. Muir had changed my whole life.

"Now, will you excuse your Mother and me for a moment? I would like to have a word with her alone," he asked. Slowly I got up, walked over to where he was sitting and put my arms around his neck lightly. All I could muster was a whisper, "I'll remember this day, and every word you said for as long as I live. I'll make you proud of me, Dr. Muir, I promise."

Then I walked out of his office into the hallway waiting area and sank into a now available chair. I must have had a strange look on my face, because nearby patients stared at me oddly, probably thinking I had just received some bad medical news. I sat in that chair feeling numb until Momma and Dr. Muir came out of his office.

"Is it okay, Momma?" I asked, looking up at her hopefully.

"Yes, Marilyn, what Dr. Muir is doing is wonderful," she replied.

Dr. Muir smiled, putting his hand on my shoulder and said, "Now you are truly ready to leave for college, young lady. Have fun and enjoy this time of your life. Write to me when you can, call if you need anything I've forgotten, and try to come by and enjoy a holiday dinner with me and my family whenever possible. I'd like that."

Wide-eyed, and not cognizant of others around me, I jumped up and threw my arms around his waist, with my head against his chest, crying as I said at least a million times, "Thank you, oh, thank you! I won't let you down, I promise. I won't let you down."

Momma and I walked out of the clinic saying goodbye to the nurses we passed along the hall. When we got outside the clinic, the rain had stopped but that fresh rain smell filled the air. I had gone into his office apprehensively and now I felt invincible, like I could accomplish anything with Dr. Muir behind me. I looked up high into the sky and thought I could feel Daddy looking down on me, with a warm smile on his face.

○ X ○ X ○

After supper, Don and I watched television with Momma and Johnny for a while. I could hardly wait to tell him about Dr. Muir's generosity, but I wanted to do it when we visited our favorite parking place on the hill one last time before we left for college.

"Can you believe it, Don? Can you believe what he is going to do for me?" I was beyond joyful. "Here I was wondering how I was going to pay for my books and dorm fees for just the first month. Now I don't have to worry about money anymore. Not only am I going to be able to go to college, but I can go without ever having to worry about money!"

Don's reaction was different than I had expected. He kept asking me, "I don't understand. Why would he do something like this for you? He has never done this for anyone else, has he?"

I couldn't answer him because I was as mystified as he was. We sat quietly that night, just holding each other, deep in thought.

Then he broke the silence. "Well, I guess it will be a long time before we park on this hill again, if ever. Won't be long before they block off the road leading up here," he said with uncanny foresight.

He looked over at me and asked, "I was wondering if we could celebrate the end of one era and the beginning of another—here in our special place. Who knows when, if ever, we will get up here again."

Caught up in the moment, I held him close and started to cry. "Oh, Don, this has been just our place, hasn't it."

○ × ○ × ○

Later, I whispered, "I hate to say it, but it's late. We better go." Don slid over to the driver's seat and started the engine. With his arm around me, we headed down the hill to Stansbury.

"No regrets?" he asked.

"No regrets," I replied.

When we got to my house, he walked me to the door. As we kissed goodnight, he said "I don't know why it is, but somehow I feel like things are going to change between us."

"Oh, Don, the only thing that is going to change is geography. Nothing else. And as for our special place, we'll be able to go there until Momma moves out of Stansbury, maybe even after that. After all the houses are moved, the road will still be there. I love you more than anything in this world.

"We'll finish our college educations and then we'll have the biggest wedding ever, that we pay for ourselves. We'll get married, have children, and grow old together. We have our whole life ahead of us." Then I added, "But, Don, if Momma doesn't remarry, we will have to help her."

"I know," Don said.

At that, I turned and went inside.

CHAPTER TWENTY-SIX
A BITTER FAREWELL

I HAD TROUBLE falling sleep the night before my move to the University. I awoke at five that morning and stared at the ceiling. I had forgotten one thing. Quietly I dressed and left Momma a note on the dining room table telling her I was going to the cemetery. I grabbed the car keys off the hook on the kitchen wall and headed for Rock Springs.

o x o x o

A heavy fog hung over the cemetery as I entered through the main gate and drove to the Muir Addition where Daddy was buried. A sense of quiet peace filled the place, like Daddy was waiting for me. I walked across the dewy grass until I reached his gravestone, knelt beside it, and spoke as if he were sitting right beside me. I told him how excited I was to be going to college and what Dr. Muir was doing for me. Then I told him about my boyfriend and what we'd done on the hill overlooking camp. I tried to justify my actions, knowing all the while Daddy would have disapproved. "You see, Daddy, after you died, I wondered which one of us would be next. I didn't want to die without experiencing real love. I hope you understand."

I looked out across the cemetery toward White Mountain and continued, "I worry about Momma and Johnny living in Stansbury all alone. It's really scary out there now that practically all the people have moved away."

When I got back home, everyone was still asleep. I tore up the note I'd left for Momma and climbed back in my bed.

o x o x o

At first university life offered something new every day: new dorm

259

roommates, rush week, pledging the Kappa Delta sorority, and having "sisters" for the first time in my life, always with Don as my anchor. I spent little time in my dorm room because something was always happening at the sorority house. I met Don at the library to study, and I called home often.

During one of my phone calls I asked Momma if she'd heard anything new about the Stansbury houses. She said she'd heard that three prominent Rock Springs businessmen, Fred Magagna, Barney Decora, and John Anselmi, had negotiated the purchase of eighty-five houses in Stansbury and Superior from the UP Coal Company, with an option to buy an additional sixty-nine Stansbury houses. The purchase price was never publicly disclosed but Anselmi, in an article in the local newspaper, termed the transaction a "good-sized deal." The buyers planned to sell some houses to buyers in Big Piney, Riverton, Dutch John (the Flaming Gorge townsite), Rock Springs White Mountain Addition, and Fort Bridger. The Winton Community Hall would become the Fort Bridger American Legion Hall. The only house not for sale was the one Mr. Bayless had given to Momma. Superior Lumber, the contractor for the new Bellview development, set a date to begin work on the infrastructure, leveling the ground, and digging the foundations.

○ ✕ ○ ✕ ○

After two weeks, Don and I caught a ride back to Reliance. As we looked out the window at the open plains, everything looked barren and gray, but for once the wind wasn't blowing. When we finally drove over the hill leading into Stansbury South Camp, I could feel myself clutch. After being away only two weeks, Stansbury looked even worse than I had remembered—bleak and rundown. The smell of mine water hung in the air. The white houses of the camp, once pristine, seemed to be covered in a brown film from the blowing dirt and sand. Almost all the houses were boarded up, with trash and tumbleweeds strewn in the yards. What was left of the white picket fences looked like broken teeth in the mouth of an old horse.

However the windowpanes of the few occupied houses still sparkled as I had remembered from days gone by. A couple of houses were jacked up from their foundations, with moving crews working around them.

There was no people on the streets, no laundry flapping on the clothes-lines, and no children's voices. Stansbury was beginning to look like a deserted town in an Old West movie.

The minute we drove into the driveway, Johnny came running out to greet us, with Tippy right on his heels. When we got out of the car, Tippy ran over to me, jumping and putting his paws on my chest as he licked my face and wagged his tail. I was so happy to see Johnny and Tippy that I couldn't stop hugging them both. I'd missed Tippy as much as I did my family.

"Are those presents for us?" Johnny asked excitedly. He chattered incessantly, telling me of all the changes since I had left, and how much fun it was to watch and help the house movers. "You should see all the money me and my friend are making!" he exclaimed.

"Money? How are you kids making money?" I asked, puzzled.

He explained that he and his friend were hired to clean out the foundations, after the houses had been removed. "People left a whole bunch of stuff in their basements, especially where the cubby hole was!" he bubbled. "We load the stuff in our wagons and the movers look it over first, then they let us keep what they don't want. They keep all the old tools, even if they are rusty, but they let us keep 'most everything else. We've sold zillions of old pop bottles and canning jars to the junk-yard in Rock Springs. The man there wants to see anything we find, and he pays us for what he wants. You should see all the neat stuff we found: old Lionel trains that had been packed away in boxes; old fishing poles and tackle boxes; boxes and boxes of old comic books; coffee cans full of pretty glass marbles and steelies. And Marilyn, I even found some dolls for you. They're dirty, but they're still pretty! One of them I found in the back of someone's toilet that was lying in their backyard."

I tried to hide my reaction to a doll salvaged from a toilet.

Johnny gushed on, "We only work after school and on the weekends while it is light outside because stealers come at night."

"Stealers? What's left to steal?"

"Oh, they steal the stuff that's too hard for us to get out, like all the galvanized steel pipes in the basements, tubs, sinks, toilets. We work one day in a foundation and the next day when we go back, all the pipes are torn out and gone. The movers told us if the men doing this ever get caught, they could go to jail, maybe even prison," Johnny said excitedly.

"Johnny! Does Momma know you're doing this?" I asked.

"She knows. Sometimes she even brings everybody something cold to drink! The movers think she's nice. I give Momma most of the money I make, and besides, it gives me and my friend something to do. The movers said we are little businessmen." He shrugged proudly.

Momma looked tired when I saw her standing in the doorway. I wrapped my arms around her and asked, "Did you miss me, Momma?"

"I missed everything about you, especially hearing your laugh," she replied. She turned and gave Don a hug, too, and coaxed him into joining us for supper.

Johnny had completely taken over my bedroom and filled it with all his toys and model airplanes, but I set my suitcases there. I went back into the living room and gave Momma and Johnny the packages I had brought for them. Johnny ripped open his package and was thrilled to see a new shirt, a pair of Levis, a package of socks and underwear, an airplane model, plus an envelope with spending money. Momma quietly opened her package, a floral cotton dress and coordinating strand of plastic beads.

"Oh, you remembered how much I love necklaces. Thanks for thinking about us." It felt good to be home with them and see them happy and smiling.

I asked if I could call Jimmy and invite him out for dinner. "You can call, but don't be disappointed if he doesn't come. He calls all the time but never comes out here because he is either at school or working."

"You mean to tell me he never comes out to see his mother?"

I was indignant as I dialed Uncle John's telephone number. The phone rang and rang, but there was no answer. I felt bad for Momma.

As we gathered around the supper table, I looked over at my little brother and thought how special he was. He never asked for much, was always eager to help everyone, and must miss having a father to put his arms around him. I wondered how much he remembered of Daddy's accident. How much was he holding inside? Once in a while, he'd ask me questions that were hard for me to answer, because I didn't have the answers myself.

Over dinner, Momma told us the house she'd chosen wouldn't be moved until October 1959, almost a year away. I hated to think about her and Johnny enduring another whole year in this ghost town. I felt so guilty living my own carefree life in Laramie while Momma and Johnny tried to survive here with Stansbury falling down around them.

The ringing of the telephone interrupted our conversation. Johnny jumped up and ran to answer it. "It's Jimmy, you guys! He's coming out," he yelled.

Fifteen minutes later, Jimmy walked into the house. After greetings and Jimmy's questions about college life, Johnny piped up and said, "I like having all you guys here like it used to be. It's sad in this house with just me and Momma."

"Oh, it's not that bad," Momma murmured as she reached down to put her arm around him. "What do you say we all take a walk around camp and look at the changes around Stansbury? Be good exercise and we could catch up on the news as we walk."

As we walked along, Momma pointed out certain things and suddenly Jimmy blurted out disgustedly, "I can't believe we ever lived in this place. Momma, it makes me sick when I think of you and Johnny still living here. There's nothing but ugly everywhere you look."

"Oh, Son, it breaks my heart to hear you talk like that. We've got to make the best of our circumstances. God put us here for a reason, and only He knows what that reason is. Let's enjoy the time we have together. It means so much to me for all of us to take this walk tonight. We probably won't have this opportunity ever again because it won't be long before Stansbury is gone." Momma had a faraway look in her eyes, as if she saw more than the rest of us.

Even though it was starting to get dark, semi-trucks still roared in and out of camp. Smaller pickup trucks parked by the empty foundations and drove around the hills surrounding the mine and the mine shop.

We made a circuit around the camp, recalling which families had lived where and special moments we'd shared with those families. Each of us reminisced about different landmarks, buildings, and even the creek that ran between north and south camps. After a while, we even had Jimmy laughing about comical events from growing up.

After Jimmy left, we sat around the television watching *Dragnet*, *Person to Person* with Edward R. Morrow, and then *The Rifleman*. I asked Momma if it would be all right if Don and I drove around for a while. Johnny said he'd watch television and wait up for me to come home before he went to bed.

Don and I drove up the winding, dirt road leading to our special place on the hill. He pulled me close and we kissed. He whispered, "I've been waiting for this moment for the last two weeks. I hate having to share you with that sorority and all those phony girls."

After silent moments, he asked, "You know, for some reason I feel something weird is going to happen to us. You still love me, don't you?"

Perplexed, I asked, "Why would you even ask such a question? Do you think I would be here right now if I wasn't crazy about you?"

After another silence filled with snuggling, he said, "I was just thinking. I can hardly wait until we are through with college because I hate it when we're at school and some other guy even looks at you. I wonder what they would think if they could see where you live and that you aren't the society girl they think you are?"

"Don, what a thing to say! I don't care about other guys and what they think. You are the one and only one I love."

○ ✕ ○ ✕ ○

Johnny was waiting up for us when we returned home. No sooner had I told Johnny goodnight and crawled into bed than Tippy ran from one room to another, whining. Momma hollered for him to be quiet. I had just begun to doze off when suddenly Tippy barked and snarled as he ran from window to window, finally jumping up on a windowsill in the front room and growling viciously. I could hear a window screen being torn from one of the windows on our house. Johnny and I ran into Momma's room. "Momma, what the hell's happening?" I shrieked out of sheer fear.

"You kids stay right here in my bedroom until I get back!" Momma instructed. "I'm going to open the front door and let Tippy out. He'll scare away whoever's out there."

Tippy lunged out the door, in pursuit of the intruders. Then we heard someone slam a vehicle door, Tippy yelp, and a truck roar off. We could hear whimpering and Momma called Tippy back to the house. He came in with blood all over him, coming from a big laceration on his head.

"Oh, Tippy!" Johnny wailed. "I'm going to kill whoever did this. Will he be all right, Momma?" he sniffled as Momma and I cleaned Tippy with towels and water. Johnny held him in his arms stroking his back, while telling him he was a good dog. Momma turned a light on in the living room before we went back to bed, but I had a hard time going to sleep for thinking of what must be going on in the darkness outside.

I called Don the next morning and told him someone had tried to steal our window screen right off our house. He offered to come spend the night with us, agreeing that it was unsafe in Stansbury right now.

The next day Momma and I drove into Rock Springs and headed for Safeway. I insisted on writing the check for the groceries, and then later I took her and Johnny to eat at a nice restaurant.

"Oh, Marilyn, you can't spend Dr. Muir's money on things for us. That money is for your schooling," Momma chided me.

"Come on, Momma. I'd have to pay for food if I were in Laramie."

When the checker totaled up the bill at the grocery, I wrote the check. We put everything into the trunk of the car and headed for the New Grand Café. Still Momma was not comfortable with me spending Dr. Muir's money on her, but I talked her into a nice meal.

Then I met Don and we went to the dance at the Eagles and visited old friends. After the dance when Don drove me home, the camp was pitch black, with only one or two dim streetlights still working. I found myself missing the sound of the mine whistle, and the muffled voices of miners walking home from the late night shift. I even missed the sound of mining going on beneath the surface, which sounded each night like it was underneath my bed. Don sat quietly with his arms wrapped around me and asked if I wanted him to stay at my house. I told him if anything happened I'd call him right after I called the police

"Marilyn! Think about it. No one came to help you last night," he reminded.

"No, Don. If you stayed tonight, I'd want you to stay every night we come home from college. This is something we have to do by ourselves."

Sunday morning I went to church with Momma to hear her sing in the choir. Momma's voice sounded beautiful at the service. I could see how she got comfort from this church, surrounded by all the people in this congregation who were her friends. After church, Momma and I stopped by Dr. Muir's house so I could visit with him for a few minutes before I went back to Laramie. He greeted me warmly, and I told him all about my first two weeks of college. He said he'd be traveling to Denver quite a bit and would stop by Laramie sometime and take me to dinner. I also told him about taking Momma and Johnny out to dinner and buying groceries for them.

"Marilyn, that's fine. Had I been in your same situation, I would have done the same thing. But thank you for telling me. Now run along. Your family is waiting." I started to walk away and then turned back. "Thank you, Dr. Muir." He smiled and kissed me on the forehead.

○ ✕ ○ ✕ ○

Don and I settled into college life. In between classes I joined sorority friends at the Union and ate my meals at the sorority house. Don sometimes joined me in the evening at the sorority house but said he never felt comfortable there. At the end of the first semester, I had the option to leave the dorm and move into the sorority.

When it was time to order a sorority pin, I called Dr. Muir and asked him how much I should spend on this special piece of jewelry. His answer was, "Get the best." I wore that pin every day while I was at the university. Don said I changed the moment I put it on, that when I wore it I had an air of arrogance. At first, I didn't take his comments seriously.

○ ✕ ○ ✕ ○

Dr. Muir had been right. The sorority opened the doors to new experiences: homecoming, athletic events, social functions. My sorority nominated me as their candidate for Sweater Queen. I got to have an oil portrait made by Ludwig Studios, which was posted all over campus, as well as in newspapers. Dr. Muir saw it in the Rock Springs paper and gave me a call. After this picture came out, my life took a different path. I started receiving calls from boys asking me to go out.

Sorority events seemed to come between Don and me more often until, in a confrontation, he asked me to choose between him and the sorority. After some tries at reconciliation, I didn't see Don anymore except when he was walking on campus or sitting with other guys in the Union. He didn't act like he wanted to speak to me. I started dating other people. I did what Dr. Muir had suggested and didn't go with just one person. One night as I was walking into my sorority house with my date, I saw Don walking a girl into the sorority house next to mine. Apparently, he wasn't moping around either.

Before I knew it, my first year of college was about to end. As I was packing my things to go home, I received a letter addressed in red ink. As I opened the letter, my thoughts turned to something Momma always said: "Never write a letter in red ink, it will bring bad luck." As I read what was written inside, I felt my whole world crash around me. Dr. Muir, who was just forty-two-years old, had died unexpectedly. His wife had written to say that under the circumstances she would no longer be able to continue the financial support of my education.

My father had been forty-two-years old when he was killed and now Dr. Muir was gone at that same age. I ran to my room and sobbed. I didn't feel like I could go on, nor did I think I could afford to return to college. In other times of tragedy I'd turned to Don, but now I had lost him too. I felt like something must be wrong with me that made my dreams turn to dust.

I ran to the phone and called Momma to tell her the devastating news. She cried, too, asking, "Now what are we going to do?"

"I'll just have to work my way through college, like I'd planned in the beginning. I'm also going to have to make as much money this summer as I can."

○ ✕ ○ ✕ ○

I didn't go back to Stansbury that summer because there were no jobs in Rock Springs. I took a waitress job at the local Greyhound bus depot and took my resumé to the Personnel Office on campus. While I waited to hear from them, I saved every penny I made. One day, a man who often came in for coffee asked me if I would like to come to work for him. "What would I be doing?" I asked.

"My name is G.R. McConnell. I'm a criminal attorney with an office around the corner."

I agreed to give him an answer the next day. After he left, I asked my boss if he knew Mr. McConnell. "He's just about the best lawyer in the state," he replied. I accepted the offer and worked days in the law office and evening at the bus depot.

I loved working at the law office. On many occasions, I went to court with Mr. McConnell, and when we walked back to the office, he liked to stop at the local drugstore and treat me to an ice cream soda. I managed to save quite a bit of money that summer.

I moved out of the sorority house and into a small off-campus apartment for the summer. Right before school started, I told Mr. McConnell that I wouldn't be able to work for him full-time anymore. He tried to encourage me not to go back to school and said he could train me to be a paralegal. "I'll tell you one thing," he explained, "you'd make a hell of a lot more money as a paralegal than you ever would as a teacher." As much as I'd miss working for him, I had to decline his offer.

Once or twice during the summer, Momma and Johnny drove down to Laramie to see me. Then one Saturday, I hitched a ride home with an upperclassman from Reliance. We went to the dance at the Eagles together that night. When I walked in, I saw Don there, dancing with a high school girl. For a few moments our eyes met, then he turned his head. We never spoke the entire evening. I experienced a feeling of sadness that what I had thought was forever had ended.

○ ✕ ○ ✕ ○

When school started in the fall, there was a big difference: I was now working my way through college. Instead of getting ready to attend sorority meetings and social functions, I was running down the back steps of the sorority house and walking thirteen blocks to wait tables at the bus stop.

I struggled financially and reached the point where I was working three part-time jobs on campus besides waiting tables, leaving very little time to study. From time to time I saw Don on campus but I never stopped to tell him about Dr. Muir. I think he knew. I missed Don, but started seeing one fraternity member, Roy, almost exclusively.

○ × ○ × ○

In October 1959, UP notified Momma that the house she had chosen had been moved to the new Bellview Addition, and they'd installed a new roof and utility hookups. The city was acting quickly to begin the infrastructure for the development.

Momma's house was the first one in the addition and sat on rugged terrain of clay dirt and slabs of rock—not a tree anywhere. The first time she and Grandpa went inside the house and saw the damage done in moving it, she realized what a daunting task lay ahead. If Momma was reluctant to embrace the challenge, she never voiced it to me. Grandpa volunteered money for supplies and taught Momma the skills she needed by working along with her to complete the repairs.

She was excited when she learned that many former Stansbury families planned to move houses they'd purchased from UP into the new addition, making it almost like recreating the "coal camp."

Johnny transferred to the Rock Springs School District, and he and Momma moved in with Grandma and Grampa. They left behind in Stansbury all of the household items until the Bellview house was finished. Thank goodness, no one ever broke into the Stansbury house.

Momma continued to work part-time in the mornings and spent the afternoons and weekends with Grandpa cleaning the dirt and debris out of the house. Grandpa's 1949 red Ford car was always parked in front of the house waiting for her to arrive. Each morning she packed a lunch for both of them.

At this time, there were no defined streets. As the snow melted, the whole area became a muddy mess. Many times they got stuck in the clay-like mud and had to be careful not to slide over the edge of the hill while getting up to the house.

They carted water from Grandpa's house day after day until finally the city hooked the water up, the first flowing water on the hill. One day a construction worker knocked on the door and asked if he could connect a hose to the outside of the house. He offered, in return, to grade the dirt around Momma's house, and she was happy to accept.

Once the utilities were hooked up, Momma and Grandpa patched the sheet rock that had cracked during the move; sanded and refinished

The Stansbury house, after it was moved to Bellview, needed many repairs, which Grampa and Momma made themselves.

the hardwood floors on their hands and knees; rehung all the doors; laid new linoleum in the kitchen and bathroom; and repainted the rooms. At the end of each day, they went back to Grandpa's house exhausted, with just enough energy to eat the hot meal Grandma had waiting for them, before they fell into bed.

Momma was so busy with the house that she had no time for any kind of social life. She never asked any of the men she knew to help with the house. In fact, most knew nothing of the house until she moved into it.

O X O X O

In November 1959, just before Thanksgiving, two years after the mine closed, Momma and Johnny were the last people to move from Stansbury. It had been a four-year ordeal after Daddy's death. One of Johnny's friends and his father moved the smaller items with their pickup truck. Then Momma found a mover who would take payments and hired him to move the big furniture. When I spoke to her on the phone I asked, "But what about your two brothers in Rock Springs, Momma? Can't they see you need help?"

"I didn't ask them. They've never offered to help before. But that's okay. We'll make out okay," she replied.

"And what about Jimmy, Momma? Couldn't he help?

"Oh, yes. But the poor kid is already working long hours at the bakery and at the White Mountain Lodge. He comes out whenever he can and hauls boxes into the house in Rock Springs for me. I had to call him because I was so tired I just couldn't lift boxes anymore. And he came out the day the mover was here and helped load the heavy furniture. Our Jimmy is a good boy," she replied.

Somehow Momma got moved into Rock Springs, and then she called to make sure that I'd be coming home early for our first Thanksgiving in the new house. She asked me to help her do some final things in the Stansbury house. She didn't like being out there by herself anymore, even during the day.

Momma was still sad that Don was no longer part of our family and told me so many times. She hadn't met Roy, but I planned to bring him home with me when school got out that spring.

When I got home for the Thanksgiving holiday, I was amazed at the job Momma and Grandpa had done in turning the house into a home. I could feel Daddy's approval as I looked around. "Momma, I can't believe how nice the house looks. If I didn't know better, I'd think a professional contractor had done the work."

"But wait a minute. Where's Tippy?" I asked. I'd just noticed that he hadn't run up to greet me.

"Oh, he's still out in Stansbury. We'll bring him back with us tomorrow when we go out there to close the house down completely," she answered.

Momma and I set a table that could have come from *Good Housekeeping.* A linen table cloth, ruby red dishes Daddy had bought for her years ago for holidays, a centerpiece, special silverware—the table looked elegant.

The smell of a roasting turkey filled the house as Jimmy, Grandma, and Grandpa arrived to join the three of us. I loved being around the table with my family, but Daddy's empty chair at the head of the table reminded us that we weren't complete.

o x o x o

After breakfast the next day, Johnny went to spend the day with friends and Momma gathered her work clothes for Stansbury. A thick cover of grey clouds, biting cold air, and moisture permeated everything. Snow was predicted.

As the car turned the final bend to approach the camp, I gasped in disbelief. All that was left was a graveyard of foundations. Like an oasis on a desert, our house stood alone on its foundation. The wind in all its fury swept and howled eerily through the rubble. There wasn't another soul anywhere 'though I knew there must be a watchman somewhere. I looked over at Momma and realized that things which once had thoroughly frightened her now slid off her back. She had lost so much of what mattered most to her that she no longer felt threatened by anything.

As we drove up to the place we'd called home for twenty years, we both started crying. "Momma," I quietly sobbed as I reached to hold her, "why did all this have to happen to us? Our family was so special."

She didn't reply to this question I'd asked her time and time again.

As I held her, I looked over her shoulder and down the street to the road leading to the special parking space on the hill Don and I had called our own. The road had been bulldozed and blocked with a huge wooded gate with a NO TRESPASSING sign. Seeing that barricaded road punctured my heart as if it was the final nail, hammering that time of my life closed.

I missed Daddy now even more than I did those first days and weeks after his death. I wondered if I would have been a different person if he been there to guide me as I grew up? A better person? Would I have made different choices? Better choices? What would my brothers have been like, had he lived? And, Momma—her life would have been so different.

"Oh, Momma," I repeated as I stepped out of the car and looked at her. "Someone even dug up the trees that were in our yard. How could people do this before we'd even moved out? It's a good thing Grandpa took the fence and grass when he did."

Momma turned and went into the house to change into her work clothes. I called out Tippy's name, looking everywhere for him. I called and whistled for him, but he didn't answer with barks of excitement like he usually did when I called him. Momma came walking out of the house with a strange look in her eye. "Momma? Where is Tippy? I've been calling and calling for him. We'd better go look for him. Where is he, Momma?" I asked.

With her head down, she said softly, "Marilyn, we're not going to go look for him. Tippy is gone. I didn't want to tell you. Your brothers and I have already looked everywhere for him, until we didn't have any more places to look. Jimmy and Johnny are heartbroken, and so am I. He was a wonderful dog."

I burst into tears all over again and sobbed inconsolably. "Oh, my God, I should never have left for college. I should have waited. You needed me so badly, Momma, and I left you when you needed me. Now Tippy's gone. Daddy's gone. Everything is gone." I ran around to the back door where Tippy's doghouse sat by the side of the porch, empty except for old blankets lining the bottom. Momma followed me.

"Come on now, Marilyn."

"I just wanted to look at his doghouse again. Do you think we should put his house in the trunk and take it to Rock Springs in case he finds us?"

"Come on," she replied. "I'll give you a hand carrying it to the car. Who knows? Maybe he will show up one of these days, or someone will find him and bring him into town for us." We each grabbed onto the roof and loaded the doghouse into the trunk without saying another word. Then Momma turned to me and solemnly said, "I don't want you to bring up Tippy again. It is just too painful for Johnny."

Momma reached over and put her arms around me, and we cried together. Then she stood back and said, "We just have to go on. Let's go back inside and get those last things out of the basement, and then we'll leave here for good."

When we entered the house it was so cold we could see our breath, since both the furnace and the water heater had been off several days. I hugged myself to keep warm. "Let's hurry and get out of here," I said, as my bottom lip began to quiver.

Together we carried the few remaining boxes in the basement to the back seat of the car and went back into the house to do a final walk through. The empty shell of the house brought on a feeling of remorse. How could Momma stand to keep saying goodbye?

We went into the basement one last time to be sure we hadn't forgotten anything. I stood on the bottom stair while Momma looked on the other side of the furnace. All of a sudden I witnessed the most horrible sight I had ever seen. Without any warning, the sewer started making a loud gurgling sound and immediately started backing up, belching up sewage from the drain in the basement floor. It rose so fast that within moments the foul smelling liquid was up to Momma's knees. I became almost petrified as I tried to think of how to stop this. Then Momma cried out in terror, "Oh, God help me." She tried to slog back toward me through the waste material all around her without losing her balance. My heart beat with such force that I could feel it rising in my throat, and I was overcome with a feeling of total helplessness. The sludge was exploding out of the drain with such force that the current threatened to knock her over. I hardly recognized the blood-curdling screams that came from me. "Momma, hurry. You'll drown in that crap if you don't hurry. I can't get to you!" She was so horrified and scared that for a moment she just stood there with the sewage rising. "Move! Reach for my hand! Hurry! Oh, Momma, hurry! Here, grab onto my hands," I screamed at the top of my lungs, hoping I could give her the courage to start moving.

Momma finally pushed through the muck, made her way around the furnace, neared the stair where I stood and clutched onto my waiting hands. I pulled her toward me as hard as I could and together we climbed up the stairs leading out of the basement. She was covered from her waist down with the evil substance that had belched out of the sewer. "Marilyn, let me go into the bathroom and rinse myself off with water in the tub."

"But Momma, you're freezing, and there's no hot water."

"I know. But I can't get in the car like this. I have to rinse myself off and take off these filthy clothes. Thank goodness, I have my dry clothes." Momma sat on the edge of the tub and grimaced as she splashed ice-cold water, rinsing herself off. She stepped out of the tub and quickly wiped herself down with an old cleaning towel that was hanging on a rack near the tub. She still smelled horrible.

She pulled the clean clothes onto her still filthy body, then collapsed down onto the lid of the toilet and sobbed. Finally through clenched jaws, she yelled, "Let's get the hell out of this place once and for all. This damn mining town! I worked my ass off in this horrible place and for what?" I'd never heard Momma use words like those before and knew she'd reached her limit.

We ran to the car and I yelled, "I'm gonna drive, Momma," as I jumped behind the steering wheel, started up the car, and turned on the heater. She compliantly got into the passenger seat. With her eyes hard, and her voice harder still, she said, "I never wanted your dad to be a miner, and he never wanted to be one either, but he had no choice. He needed a job! Then he came to love that damn mine until it finally killed him. I tried and tried, but I just couldn't get him to leave, Marilyn. I just couldn't get him to leave. When we drive out of this camp today, I never want to come back, ever! Do you hear me, Marilyn? Our roots here are gone. I have nothing but a lot of bad memories in this place. It was such a hard life and for what? In the end, it took everything I loved." I listened to all she said without making a sound, tears filling my eyes until I could hardly see the road ahead. I wished she'd stop, as her words were tearing my heart out, but somehow I knew she had to let the bile out to save herself.

When I reached Rock Springs, I drove straight to Grandma and Grandpa's house to tell them what had happened. Momma, who said she wasn't feeling bad but looked terrible, now sat listlessly in the seat. I ran in to get Grandpa who came out, took one look at her, and said in broken English, "I go call Dr. Pryich." I waited in the car with Momma, watching her breathe without either of us saying a word. He came running out of the house and said, "Mala, you take your mother straight up to the clinic. Doctor will be waiting for you. You take her straight into his office. No sitting in waiting room."

Dr. Pryich had grown up in the neighborhood, and he'd taken Dr. Muir's place at the clinic. The receptionist wasted little time in getting us in to see Dr. Pryich. He took one look at Momma, checked her over thoroughly, and then looked at me and said as he gave me a packet of antibiotic pills, "Marilyn, your mother is a very sick woman. There's no telling what toxins were in that sewage. Don't stay in your house tonight.

Stay at your grandparent's house with her until she gets over this. Put her to bed, keep her warm, and wake her to be sure she takes one of these pills every six hours. I'll call and check on her after I do my rounds at the hospital tonight. This antibiotic is very expensive but it's a good one. I've given you enough samples to see her through this."

I did as he said. Around ten that evening, the doctor knocked at my grandparent's door. We were all so glad to see him that Grandpa greeted him with, "Praise the Lord." Momma was in a deep sleep, her hair wet from fever. The doctor checked all her vital signs and then asked for a pan of cool water and a washcloth, which he used to moisten her face. While he did this, Momma opened her eyes and looked up at him. "Oh, Dr. Pryich, I don't know when I've been this sick."

"I know, Margaret. I know. But you're going to be okay. If this medication doesn't break the fever, I'll have to admit you to the hospital. It's going to take some time for you to get over this. Just keep taking the medication."

After a while the doctor left, and I cuddled up in a blanket on Grandma's favorite chintz-covered chair that she kept by the side of her bed. The room was cold and silent. The only sound was the howling wind outside as the chill seeped in through the bedroom window.

When the alarm went off in six hours, it took me awhile to wake Momma up to give her the pill. Then she fell right back into a deep sleep. She looked so vulnerable and alone lying in the bed. She needed Daddy's arms around her now more than ever. She had worked so hard with everything involved in the move and now it seemed the good Lord was forcing her to rest. Momma was in bed for three straight days, awakening only long enough to take the medication. Dr. Pryich telephoned every morning to see how she was doing, even after the fever broke. All of us were so worried we were going to lose her, but especially me and my brothers. She was all we had left. Since Daddy had died, she had repeatedly demonstrated strength and courage and adapted to circumstances that would have defeated most people.

○ ✗ ○ ✗ ○

Experienced miners explained to me later that the event Momma endured that day wasn't an ordinary sewer problem. The mine always had water in it, and UP continually ran pumps to flush the water out of the mine into the sewage system—the same system the houses were connected to. When the mine closed, UP turned the pumps off and the mine filled with water, creating more and more pressure on the sewer system.

That day, the pressure must have mounted until it reached the limit and the mine water and sewage burst loose, gushing downhill through the pipes until it reached an open drain where it erupted out. Our house was the only house left in town, and Momma and I just happened to be in the basement that day. They said that the whole basement probably eventually filled with the evil-smelling sludge until it ran out the windows and down the street.

I stayed with Momma all that week until she was well enough to go back to our new house. To explain my absence from school, I called the dean of women, the sorority housemother, and my part-time employers. When Momma was better, I took the train back to Laramie and caught a cab to the sorority house. Even after Momma went home, it took weeks before she regained her strength and resumed her daily routine.

I called Momma every night to make sure she didn't need me. I didn't go home for the next couple of weekends as I needed to catch up with my schoolwork and my part-time jobs. Roy helped me when he could and insisted that I not dwell on circumstances back home. To cheer me up, we went to my first college basketball game. Our team had a young player recruited from New York who played in the guard position. When he came onto the court, the crowd went wild, so I knew he must be good. He had a unique way of shooting with both hands over his head, hitting the basket from anywhere on the court. Roy and I never missed a home game after that night. During one of the games, Roy became annoyed at my zealous behavior and looked at me and said, "If you like him so much, maybe you should marry him."

"Why would you even say something like that? I don't even know him. I just admire his athletic ability, like everybody else, but I'm pinned to you." I moved my hand to his fraternity pin on my sweater, a symbol of pre-engagement.

o x o x o

Finally Momma returned to work and got completely settled in the house. Our lives went on. Throughout all this time, I never stopped thinking about Don. I wondered if he, too, had met someone to love, how he was doing in school, how his parents were, and if he ever thought of me and the love we once shared at our special place on the hill.

CHAPTER TWENTY-SEVEN

SUNSET OVER STANSBURY

DAY BY DAY, Stansbury wore away like an old shoe. The town sign that the 4-H Club members had worked so painstakingly to construct was gone completely. Vandals left only the splintered four-by-four post that had once supported it.

The only traces of what had once been a thriving mining community were the crumbled foundations of houses. The camp looked like a bombed out war zone once seen in Germany. Even then, people continued to scavenge the plains for any remaining discarded items left by the families. After a while, there was no longer any reason for anyone to go there, and Stansbury was left to rest in peace.

○ × ○ × ○

A short time after Momma moved into Rock Springs, a man appeared at her door. He quickly introduced himself and explained that he had worked on the crew moving houses from Stansbury. He'd heard of Momma and Johnny's frantic search for Tippy and had come to tell her what had happened. He had found the remains of a dog in an empty house his crew was preparing to move. He speculated: "A couple of young boys used to run through those empty houses with that black and white collie always with 'em. From what I can tell, the dog must have been with them one day when the wind was blowing hard, and it must have slammed the door closed behind them, trapping the dog in the house and making it impossible to hear his barking. I would have come to you sooner, but I just now learned where you had moved after you left the camp."

"What did you do with his body?" Momma asked. Tears welled in her eyes, and the man blew his nose loudly.

The Stansbury Boys' 4-H Club made this sign marking the turnoff to Stansbury. Front: Jerry Lewis, Jerry Hereford, Jimmy Nesbit, Johnny Nesbit, Unknown. Back: Bob Kalishinsky, Bob Brown, George Pryde, Unknown, Ronald Robinson, Ronnie Henderson.

"I'd heard what happened to your husband, ma'am, so, we buried the dog to the right of the mine portal. Thought that would be a fitting place, under the circumstances. I made a small cross out of some scrap wood and put it there for him."

"That was so kind of you to take the time to do that. Thank you for going out of your way to come and tell me. The kids and I looked everywhere for that dog, but never thought to look inside the empty houses. I'm glad to know what happened, but I don't think I'll take the kids out there. It would just be too painful for them to go through that all over again."

○ × ○ × ○

The following summer, rumors began to spread around Rock Springs that hippie teenagers were living in the vacant foundations in Stansbury. The police were often called there to investigate, sometimes sent by parents of missing children from all over the country. Because of the liability issue, the Union Pacific had no choice but to go one step further. A notice appeared in the local newspaper announcing the resolution and when it would happen. Word spread quickly.

I heard that one morning as the sun was coming up and the air was still, bulldozers headed for Stansbury to cover over the entire town. Lining the main road leading to the camp were dozens of parked cars belonging to miners who had once worked at Stansbury and wanted to observe Stansbury's final chapter. Onlookers stood beside their vehicles or sat motionless inside them, making no attempt to engage in conversation. Soon the steel blades were lowered and the grading project began obliterating the final remains of what once had been a thriving, productive coal mining community.

After a while, only a few trees had been spared the blade. The town where the mine had once offered men good paying jobs, meat on the table, roofs over their heads, and presents under the tree was gone. Only the dark world below, rich in coal, was left intact to be explored another day.

One by one, Momma, Jimmy, Johnny, and I left the Rock Springs area. We were able to move on to different ways of life. If there truly is a Supreme Being and our loved ones do look down on us, I think Daddy would agree we had made it against all odds. Or, maybe it was he who guided us all along the way. We had kept our family together.

○ ✕ ○ ✕ ○

I got married and left college after my sophomore year, my brothers graduated from Rock Springs High School, and Momma continued working different jobs in Rock Springs. Then, in 1965, the Vietnam War accelerated.

Jimmy, who knew he would be one of the first ones drafted, quickly signed up to join the Marines for a three-year hitch. He wanted the opportunity to specialize in electronics and possibly avoid being sent to Vietnam. When he told Momma that he had signed up, she immediately began boxing up all her personal belongings to move, without knowing where she was going. She felt Johnny would be called up next, and then she would have no reason to stay in Rock Springs. She put an ad in the local newspaper to rent out her house.

Within a few weeks, Johnny was drafted by the Army for a two-year hitch. Jimmy went to the Marine training base at Camp Pendleton, and Johnny to the Army training base at Fort Ord in California.

For each farewell trip, Grandpa pulled a fifth of bourbon and two shot glasses out of a sack he carried. In the old country tradition, with tears in his eyes, he poured a small amount of liquor in each glass and had a farewell toast with each of my brothers saying, "Salute. Keep your head down."

After the boys left, Momma rented out her house and moved to Denver to live with Aunt Ann while she looked for work. For a while she worked as a secretary in Uncle Zeke's wholesale liquor business. Then she passed the civil service exam and worked in the purchasing department with the State of Colorado, later transferring to the inheritance department. During the trona boom in Rock Springs, she sold her house for a very good price. The money she received from the sale of her house, her job, and the monthly checks from Daddy's black lung benefits allowed her to get her own place in Colorado. She remained in Denver until she retired, then moved to Phoenix to be near Johnny who bought her "her dream house" in a retirement community where she could enjoy year-round nice weather. She never remarried and passed away when she was ninety-three-years old.

After Jimmy returned from the Marines, he became vice-president of a savings and loan company in San Francisco and bought, renovated, and then sold older homes there. He never married and died at age fifty.

When Johnny came home from the Army, he graduated from Arizona State University with a degree in advertising and marketing, minoring in education. He held corporate positions with Curtis Mathis and RCA. He married his high school sweetheart and together they adopted two children, Jason and Jamie. Later they divorced and he remarried. He retired from Qwest as an advertising consultant.

Don and Roy each obtained engineering degrees and went on to marry other people. As for me, I never went back to college or became a teacher. Who would have guessed I would end up marrying an Italian boy from New York, Tony Windis—that same University of Wyoming basketball player I cheered for at games, who went on to play professionally for the Detroit Pistons and later was inducted into the University of Wyoming Athletic Hall of Fame. He returned to Wyoming where he became the first Wyoming high school coach in history to win both a boys' and girls' state basketball championship in the same year. We had two children, Anthony and Deborah. We later divorced, and I married a man born in England who was an electrical foreman at the trona company outside of Green River, Wyoming, where I worked as a secretary. We moved to Kellogg, Idaho, where he worked for the Bunker Hill Mining Company. We also owned and operated Silver Valley Electric and a convenience store in Kellogg, Idaho. We had one child, John. Later we divorced, and John and I moved back to Laramie.

Ironically, the predictions of the fortune-teller Scoot who read my palms so long ago came true: I married two times and had three children.

O X O X O

Our life in Stansbury and, in particular, the night of the accident, was something Momma, my brothers, and I carried with us throughout our lives. Over the years, my brother Johnny and I visited Rock Springs and on each occasion, we drove out to Stansbury even though there was really nothing left to see, just the geography. We'd walk in silence over the plains remembering our experiences there. On one occasion, as my brother and I were leaving, Johnny laid two fingers to his lips and whistled loudly, just like Daddy did years ago when it was time for us to come home. For one split second, at the sound of that whistle, Daddy was back with us.

In recent years, miners who had worked for the Union Pacific Coal Mines in Sweetwater County return to Rock Springs for one weekend in July to attend the annual coal camp reunion. Miners and their families come from all over the country to see and reminisce with old neighbors and friends, revisit the camps, and dance to polka music. But Momma held true to her word. She made many trips to Rock Springs to see her mother and father, but she never attended even one coal camp reunion or drove to Stansbury again.

O X O X O

Today all that remains of Stansbury is a vast deposit of bituminous coal far beneath the surface, awaiting the next demand.

A friend told me that decades later deep in the mine near the entrance to the number seven seam she saw a sign made of old mine wood nailed to the wall. Painstakingly carved into the wood was:

John Nesbit, November 10, 1955.

RESIDENTS KNOWN TO HAVE LIVED IN STANSBURY

(Please notify the author of additional names or corrections at her email address: mjw@uwyo.edu.

Children's names in parenthesis.
Aguarri, Atillo & Mary (JoAnn)
Ashby, Eugene & Velma (Eugene Jr.)
Baca, Joe & ? (Ross, Josephine, Edward, Joey)
Bergandi, Richard & ? (Richard, Rosie)
Berlich, Frank & Kathryn (Mary)
Besso, Ernest & ? (Rosa, Richard)
Bobchalk, George & ? (Robert, Leon)
Bollinger, Robert & Martha (Donnie, Patrick, ?)
Bozner, Joe & Rosilia (Dolly, Joey)
Bozner, John & Evelyn (Jonelle, Patty, Johnny)
Branstetter, Carl & ? (Dixie, Carl)
Bullard, Claude & Genevieve (?)
Brown, Bob and Maxine (Marilyn, Bob, Clayton)
Brown, Alfred & Lolita (Maxine, Patricia, ?)
Caller, Joe & Audrey (Darlene, ?)
Chetterbach. Louis & Margaret (Louie, Donna, Bill, ?)
Clinkenbeard, ? & ? (Barbara, Peggy, Mary Ann)
Coet, ? & ? (Beverly, David)
Coleman, ? & Bernice (James, Phyllis, Kendall)
Croneym, Mike & ? (Michael)
Cummings, James & Ann (Bill, JoAnn)
Daniels, Clyde & Florence (Patsy, Audie, Lottie)
Deleno, ? & Dorothy, (?)
Delgado, Frank & ? (?)
Dernovich, Tony & Rose (August, Linda)
Dompson, William & Bert (Billie, Carol, Tommy)
Duncan, ? & Thelma (?)
Eccli, Primo & ? (Bonnie, Primo)
Edlund, Alert & Janet (Loranna, Maurice)
Egan, ? & ? (Barbara)

Evans, Mike & Edith (Pauline)
Evans, David & Dale (?)
Fabiny, George & Gloria (Darlene, Gail Ann, Georgie)
Flipovich, Mike & Leona (Carolyn, Marilyn, John)
Forningo, John & Madge (?)
Fortuna, John & ? (Ray, John, Mary Ann)
Frapport, ? & Mary (Helen, Rosemary)
Galinski, Eggie & Eva (Geraldine, Bobby)
Ganger, ? & Reggie (?)
Ganzeland, ? & ? (?)
Goddard, Walter & Lucille (LaRue, Connie)
Goich, Sam & Mary (Sam, Mary, Eli, Louis, Melvin, Mary, Shirley)
Grosso, Charles & Velma (Sandra Kay)
Hanley, William and Dovey (?)
Head, Robert & Irene (Robert, Carol)
Henderson, Bob & Helen (Bob, Ron, Don, Leta Jane)
Hensley, Jack & Lena (Karlene, Jack)
Herefore, J.D. & Pauline (Nora Lee, Jerry)
Jenkins, Ray & Velma (Kay, Sherrie)
Jenkins, Thomas & Elsie (Carolyn)
Kalasinsky, Walter & ? (Martha, Robert)
Kalinosky, Victor & Gertie (Carol, Shirley, Lila)
Keywatz, ? & ? (Burt, Janette, Unice)
Kobler, ? & Margery (?)
Kragovich, Paul & Velma (Carmen, Norma Jean)
Kurtzman, ? & Elsie (?)
Lacy, Jean
Lanoy, Leon & Pauline
Law, Jim & ? (Sharon)
Lewis, William & Mildred (Billie, Jerry)
Lee, James & Zoe (Zoe, Bill, Dick)
Logan, Joe & Bernice (Sharron, Connie, Joey)
Maffoni, John & Mary (John [Meach], Jeanne, Alma)
Manzanares, ? & ? (Rosalee, Phil, Bobbie)
Marinoff, Pete & Bertha (?)
Matson, Raino & Enid (Carol, Bobby)
Mecca, Jim & Mae (Lois, Jim, Ernest)
Mecca, Pasco & Evelyn (?)
Miller, Joe & Catherine (Jack, Bill)
Miller, ? & Shirley (?)
Moser, Red & ? (Ralph, Judy)

Neal, ? & ? (Billy, Gretchen, Rosemary)
Neal, Jerry & Ruth (Geraldine)
Nelson, Carl & Octavia (Carl, Tony, Mollie)
Nesbit, John & Margaret (Marilyn, Jimmy, Johnny)
Nesbit, Bob & Arvella (Mary Jane, Barbara Kay)
Nichols, Clyde & ? (Don, Joan, ?)
Nielson, Slim & Lola
Nyquist, Myron & ? (Art)
Overy, Thomas & Axelina (Richard, Thomas, Ronald)
Palazari, Vedisto & Alice (Geraldine & Paul Guigli)
Palcher, Tony & Ladene (James, Evelyn, Carrie, Emma,
        Josephine, Helen, Lillian, Anna, Billy)
Pecolar, George & Norma (Carol, Georganne, Betty Kay)
Phillips, ? & Dolly (Bonnie)
Pryde, George & Judy (George, Coralee)
Rafferty, Jack & ? (Jack, Jim)
Ramer, ? & Marjorie (Robert)
Renz, Charles & Eloise (Charles, Dennis)
Robinson, Ron & Aurora (Ronald, Bruce)
Rogers, Joe & Joan (?)
Seppie, Albert & Lillian (Sandy, Marlene, Albert, Jimmy)
Shalatta, Bill & ? (Anna)
Sharp, Melvin & ? (Mary Elnore, Peggy)
Shultz, Minnie (Lola)
Sturman, Martin & Olin (Hope)
Swartz, Kenneth & ? (Beverly, Kenny)
Tarufelli, Geno & Martha (Eugene, Robert, Judy, Tony, Ernie, Larry,
        Jeanette, Dorothy, Linda,Tammy, Debbie)
Taucher, Urban & Mary (Patti, Urban)
Tomassi, Louis & Emma (Vincent, Anthony, LuAnn)
Uram, Leonard & ? (?)
Valco, John & ? (Jane Ann, Coralee, Johnny)
Vercimak, Mike & Mildred (?)
Volsic, ? & Phyllis (Bob, Bill)
Watson, Jim & Hazel (Jimmy, Janet)
Williamson, A.D. & ? (Bobby, Patricia, Brenda, Norman)

Beaky & Bennie Pryich were the butchers at the Stansbury Company Store.

One of the last photos of Johnny and Margaret Nesbit's three children together: Johnny, Marilyn, and Jimmy.

Marilyn Wood, who lives in Laramie, Wyoming, retired from the University of Wyoming after twenty-four years of working for head football coaches Pat Dye, Al Kincaid, and Dennis Erickson, and in the College of Agriculture. She cherishes the time she spends with her three children, Anthony Windis, Debbie Buchhammer, and Johnny Wood, and four grandchildren, Anna Marie, Thomas, and James Buchhammer and Jordan Wood.

She loves music and enjoys playing the organ, knitting and quilting, reading, corresponding /visiting with Amish friends, traveling, and swimming with friends at the recreation center. She has never lost her love for dolls and enjoys searching antique stores for dolls of the 1950s to add to her collection.

Friends can reach her at mjw@uwyo.edu.

❦ NOTES ON THE PRODUCTION OF THE BOOK ❦

The text is set in type from the Adobe Garamond family.
Display type is Anodyne from the Yellow Design Studio.
Additional fonts are Journal Hand from Typadelic
and Tessaroni Bold from the Coyote Foundry.

The text is printed on sixty-pound Joy White,
a white, acid-free, recycled paper.
The book is covered with ten-point stock,
printed in four colors, and coated with matte film lamination,
Thomson Shore.